T0396144

MEMBRANE SEPARATION PROCESSES

MEMBRANE SEPARATION PROCESSES

Theories, Problems, and Solutions

AHMAD FAUZI ISMAIL
Advanced Membrane Technology Research Centre (AMTEC), Universiti Teknologi Malaysia, Johor Bahru, Johor, Malaysia

TAKESHI MATSUURA
Department of Chemical and Biological Engineering, University of Ottawa, Ottawa, ON, Canada

ELSEVIER

Elsevier
Radarweg 29, PO Box 211, 1000 AE Amsterdam, Netherlands
The Boulevard, Langford Lane, Kidlington, Oxford OX5 1GB, United Kingdom
50 Hampshire Street, 5th Floor, Cambridge, MA 02139, United States

British Library Cataloguing-in-Publication Data
A catalogue record for this book is available from the British Library

Library of Congress Cataloging-in-Publication Data
A catalog record for this book is available from the Library of Congress

ISBN: 978-0-12-819626-7

For Information on all Elsevier publications
visit our website at https://www.elsevier.com/books-and-journals

Publisher: Susan Dennis
Acquisitions Editor: Anita Koch
Editorial Project Manager: Hilary Carr
Production Project Manager: Sruthi Satheesh
Cover Designer: Miles Hitchen

Typeset by MPS Limited, Chennai, India

Contents

Preface ix

1 Solubility parameter 1
1.1 Why is the solubility parameter necessary? 1
1.2 Theory 1
1.3 Examples of the solubility parameter 3
Nomenclature 15
References 16

2 Triangular phase diagram 17
2.1 Principles 17
2.1.1 Thermodynamics of polymer solution 17
2.1.2 Solutions for the ternary system using Flory–Huggins equations 23
Nomenclature 30
References 30

3 Reverse osmosis, forward osmosis, and pressure-retarded osmosis 31
3.1 Reverse osmosis 31
3.1.1 Reverse osmosis performance 31
3.2 Concentration polarization 35
3.3 Prediction of RO performance considering concentration polarization 37
3.4 Pore models 42
3.4.1 Preferential sorption-capillary flow model 42
3.4.2 Glückauf model 45
3.5 Forward osmosis and pressure-retarded osmosis 49
3.5.1 Principles of forward osmosis, reverse osmosis, and pressure-retarded osmosis 49
3.5.2 Applications of forward osmosis 50
3.5.3 Concentration polarization in forward osmosis 51
3.5.4 Forward osmosis transport 52
Nomenclature 59
References 60

4 Nanofiltration **61**
4.1 Solution in general 61
4.2 Solution for mono-monovalent electrolytes 63
4.3 Solution method 63
Nomenclature 67
References 68

5 Ultrafiltration and microfiltration **69**
5.1 Ultrafiltration: gel model 69
5.2 Microfiltration: Brownian diffusion, shear-induced diffusion, inertia lift 71
5.2.1 Brownian diffusion 71
5.2.2 Shear-induced diffusion 73
5.2.3 Inertial lift 73
Nomenclature 74
References 75

6 Membrane gas separation **77**
6.1 Solution-diffusion model 77
6.1.1 Steady-state transport 77
6.1.2 Unsteady-state evaluation of S and D by the time-lag method 79
6.1.3 Separation of binary gas mixture 84
6.1.4 Resistance model 86
6.2 Gas transport in porous membrane 96
6.2.1 Transport mechanism 96
6.2.2 Separation of gas mixture by the porous membrane 98
6.2.3 Measurement of pore size and pore size distribution 102
6.3 Mixed matrix membrane 105
Nomenclature 109
References 110

7 Pervaporation **113**
7.1 Pervaporation transport 113
7.2 Pervaporation transport model by Greenlaw and coworkers 121
7.3 A new model for pervaporation transport 125
Nomenclature 131
References 131

8 Membrane distillation 133
8.1 About membrane distillation 133
8.1.1 Process principles 133
8.1.2 Different membrane distillation configurations 134
8.2 Transport in direct contact membrane distillation 135
8.2.1 Heat transfer 135
8.2.2 Mass transfer 138
Nomenclature 144
References 145

9 Membrane contactor (membrane absorption) and membrane adsorption 147
9.1 Membrane contactor 147
9.1.1 Transport in membrane contactor 149
9.1.2 Wilson plot 152
9.2 Membrane adsorption 155
9.2.1 Membrane adsorption process outline 155
9.2.2 Carman–Kozeny equation for flux calculation 157
9.2.3 Mass balance in membrane adsorption 157
Nomenclature 162
References 163

10 Membrane module 165
10.1 Reverse osmosis 165
10.1.1 Reverse osmosis hollow fiber module 165
10.1.2 Reverse osmosis flat-sheet membrane 176
10.2 Gas separation 183
Nomenclature 190
References 191

11 Membrane system 193
11.1 Two flow types 193
11.1.1 Cross-flow types 193
11.1.2 Cascade and recycle 194
11.1.3 Hybrid systems 195
11.2 Reverse osmosis systems 196
11.2.1 Reverse osmosis–nanofiltration cascade 196
11.2.2 Reverse osmosis recycle 200

11.3 Gas separator systems....203
11.3.1 Gas separator cascade....203
11.3.2 Gas separator recycle....206
11.4 Forward osmosis–reverse osmosis hybrid system....210
11.5 Membrane reactor....215
11.5.1 Description of the membrane bioreactor....216
11.5.2 Bioreactor modeling....218
Nomenclature....228
References....231

12 Cost of water....233
12.1 Calculation of water production cost by Desalination Economic Evaluation Program 2000....233
12.1.1 Case identification and site characteristics box....235
12.1.2 Technical parameter input data....235
12.1.3 Membrane water performance data....235
12.1.4 Economic parameters input data....236
12.1.5 Reverse osmosis plant cost input data....237
12.1.6 Single-purpose plant performance....238
12.1.7 Stand-alone reverse osmosis water plant performance....238
12.1.8 Stand-alone reverse osmosis water plant costs....242
12.1.9 Power plant....245
12.1.10 Stand-alone reverse osmosis plant....245
References....246

Appendix....249
Index....261

Preface

In the last few years of our time at university, we used to visit several chemical plants. The chosen chemical plants were not necessarily the modern fully automated ones, where the students could only enter a room with control panels or look at the towers from remote distances. Instead, we visited rather old plants, where we were allowed to enter the area where we could watch, smell, taste (if allowed), hear the noise, and feel the temperature to understand the details of the chemical operation.

This book was written in the spirit of conventional chemical engineering education. All problems can be solved not by using sophisticated and ready-made computer software but by using a simple handheld calculator or Excel program with mathematics at an early undergraduate student level. Thus the problems are solved not by the black-box approach but by understanding the details of the computational process.

As the authors also attempted to include as many membrane separation processes as possible, this book consists of the following 12 chapters.

Solubility parameters are useful to understand the interaction between the feed components and membrane as well as to determine which components will preferentially permeate through the membrane. Solubility parameters provide important information about which solute in the feed solution is adsorbed to or rejected from the membrane surface. Therefore, in Chapter 1, Solubility Parameter, discussions take place on the solubility parameter and some example calculations are shown.

In the wet phase inversion process, the polymer solution is cast into a film, which is then immersed into a gelation media that is nonsolvent for the polymer. While the film is in the gelation media, solvent–nonsolvent exchange takes place and the polymer solution composition crosses the phase boundary line, at which point the solution is split into polymer-rich and polymer-lean phases. In Chapter 2, Triangular Phase Diagram, examples are given to guide readers in drawing phase boundary lines on a triangular diagram.

When the aqueous solutions of two different salt concentrations are separated by a semipermeable membrane, which allows the permeation of water but does not allow salt permeation, there is a natural tendency for water to flow from the solution of the lower concentration to the solution of the higher concentration. The driving force for the water flow is the difference in osmotic pressure. Forward osmosis (FO), reverse osmosis (RO), and pressure-retarded osmosis (PRO) processes are all based on this underlying principle of osmosis. In Chapter 3, Reverse Osmosis, Forward Osmosis, and Pressure-Retarded Osmosis, the fundamental transport theory is presented for FO, RO, and PRO. An attempt is made to predict the RO performance, assuming the presence of pores.

The pore size of nanofiltration (NF) membranes is slightly larger than RO membranes, ranging from 1 to 10 nm. Since the membrane is often charged, the effect of the membrane charge should be considered in the transport theory. Chapter 4, Nanofiltration, shows how NF performance can be predicted using the Nernst–Planck equation in which the steric effect and the effect of the electrostatic interaction are combined.

In Chapter 5, Ultrafiltration and Microfiltration, the gel model is introduced for ultrafiltration (UF) to explain the leveling-off of the flux with an increase in pressure. The flux of microfiltration (MF) membrane can be explained by Brownian diffusion when the particle size is smaller than about 1 μm. However, when the particle size is larger than 1 μm, the experimental values become orders of magnitude higher than the theoretical ones, which is called the "flux paradox for colloidal suspension." In order to overcome this contradiction, shear-induced diffusion and inertial lift are introduced. Some example calculations are made for both UF and MF.

Chapter 6, Membrane Gas Separation, discusses membrane gas separation. The solution–diffusion model, in which permeability is given by the product of solubility and diffusivity, is considered for the transport of gas in the nonporous membrane. First, the time-lag method is introduced to evaluate the solubility and diffusivity of the membrane separately. Then, the resistance model is used to calculate the overall permeability and selectivity for the multilayer gas separation membranes. As for the porous membrane, gas transport is classified into (1) Knudsen flow, (2) viscous flow, and (3) the combination of Knudsen and viscous flow, depending on the Knudsen number. The transport equations for mixed matrix membranes are also shown.

Chapter 7, Pervaporation, discusses pervaporation, where the feed liquid on the upstream side comes out from the downstream side as vaporous permeate. Transport through the pervaporation membrane is also explained by the solution–diffusion model, assuming that the pressure is constant, while solubility and diffusivity change considerably across the membrane since the degree of polymer swelling changes from the wet upstream side to the dry downstream side. Some example calculations are made, together with a newly proposed model.

Chapter 8, Membrane Distillation, deals with membrane distillation (MD). The transport of vapor in the dry MD membrane pore consists of (1) evaporation of water at the warm feed side, (2) migration of vapor through the dry membrane pore, and (3) condensation of vapor on the permeate side. Simultaneous heat and mass transfer occur in MD for various applications by different MD configurations, such as direct contact MD (DCMD), sweeping gas MD (SGMD), vacuum MD (VMD), and air gap MD (AGMD). The MD transport mechanism is either Knudsen diffusion or ordinary diffusion, or a combination thereof, depending on the Knudsen number. An example calculation is shown for DCMD.

Chapter 9, Membrane Contactor (Membrane Absorption) and Membrane Adsorption, describes membrane contactors and membrane adsorption. In a membrane contactor, the membrane pore is kept dry even though one side of the pore is in contact with liquid. The gas permeates through the pore until it is absorbed by the absorbent liquid. Unlike MD, heat transfer does not need to be considered. In membrane adsorption, the membrane acts as an adsorbent with functional groups that provide adsorption sites. Thus membrane adsorption combines filtration and adsorption. The transport is explained by a combination of the Kozeny–Carman equation and the first-order adsorption kinetics.

Chapter 10, Membrane Module, discusses the membrane module. Spiral wound and hollow fiber modules are used for RO and gas separation, respectively, with some exceptions. In the spiral wound module, the membrane envelope together with spacers is wound around the central collection tube. In the hollow fiber module, the bundle of hollow fibers is loaded in the limited space of a module. The concentration, flow rate, and pressure change as liquid or gas advances from the module inlet to the module outlet, even after the steady state is reached. Thus the simulation involves the solution of simultaneous differential equations. In this chapter, examples are given for the hollow fiber and spiral wound RO modules and the hollow fiber modules for gas separation.

In Chapter 11, Membrane System, simulations are made for the performance of the systems where permeate or retentate from the module is recycled to the feed, or several modules are connected in cascade. The modules for different membrane processes can also be combined to form a hybrid system. The membrane module can even be used as a membrane bioreactor. In this chapter, simulations are shown for RO-NF cascade, RO recycle, gas separation cascade, and gas separation recycle. As an example of the hybrid system, the combination of FO and RO is presented. Also, a simulation is made of a bioreactor where the yeast is immobilized between UF and RO membranes.

The evaluation of energy consumption and analysis of the product cost are the major concerns for the membrane process design. In Chapter 12, Cost of Water, the cost evaluation of water produced by a stand-alone RO plant is made according to the Desalination Economic Evaluation Program (DEEP) of the International Atomic Energy Agency (IAEA).

The book was written for undergraduate and graduate students, professors, and also for researchers in universities, research institutions, and industries who are learning about membrane separation processes. It is, therefore, the authors' wish to contribute to the further development of membrane science and technology in the future.

March 6, 2021

Ahmad Fauzi Ismail and Takeshi Matsuura

1

Solubility parameter

1.1 Why is the solubility parameter necessary?

The solubility parameter is useful to find a solvent or solvent mixture that can dissolve a polymer. It is also useful to know the interaction force working between the feed components and a membrane and to determine which component will preferentially permeate through the membrane.

1.2 Theory

The mixing of two components, 1 and 2, is accompanied by a free energy change

$$\Delta G = \Delta H - T\Delta S \tag{1.1}$$

where ΔH (J/mol), T (K), and ΔS (J/mol K) are the heat of mixing, absolute temperature, and entropy of mixing, respectively. Since the mixing of two components involves a large change in entropy, whether ΔG is positive or negative depends largely on the magnitude of ΔH. Hence, several methods have been proposed to evaluate ΔH. Among those, the following Hildebrand's equation (Hildebrand and Scott, 1950) is by far the most popular:

$$\Delta H_M = V_M \left[\left(\frac{\Delta E_1}{V_1} \right)^{1/2} - \left(\frac{\Delta E_2}{V_2} \right)^{1/2} \right]^2 \varphi_1 \varphi_2 \tag{1.2}$$

where ΔH_M is the total heat of mixing (J/mol), V_M is the total molar volume of the mixture (m^3/mol), ΔE is the heat of vaporization (J/mol), V is the molar volume (m^3/mol), φ is the volume fraction, and subscripts 1 and 2 represent components 1 and 2 of the mixture, respectively. ΔE is also called the cohesive energy. A physical interpretation of ΔE is the degree of attraction between molecules in a liquid. $\Delta E/V$ in Eq. (1.2) is equal to the density of heat of vaporization and is called the

Membrane Separation Processes. DOI: https://doi.org/10.1016/B978-0-12-819626-7.00002-8

"internal pressure" or "cohesive energy density" of the substance. Rearrangement of the above equation yields

$$\frac{\Delta H_M}{V_M \varphi_1 \varphi_2} = \left[\left(\frac{\Delta E_1}{V_1} \right)^{1/2} - \left(\frac{\Delta E_2}{V_2} \right)^{1/2} \right]^2 \tag{1.3}$$

It is obvious from Eq. (1.3) that the heat of mixing ΔH_M is always positive and becomes smaller as the cohesive energy densities of components 1 and 2 become closer, favoring the mixing of components 1 and 2. The quantity $\sqrt{\Delta E/V}$ is called the "solubility parameter" and can be obtained for many organic compounds from their structural formula (Hildebrand and Scott, 1950; Small, 1953). The same rule applies for polymers. The solubility parameter of a polymer can be obtained from the formula of the polymer repeat unit. If the solubility parameters of a polymer and solvent are close to each other, the polymer is readily miscible (soluble) in the solvent.

The heat of vaporization can be divided into three components, with each component representing a molecular interaction force of different kinds, that is,

$$\frac{\Delta E}{V} = \frac{\Delta E_d}{V} + \frac{\Delta E_p}{V} + \frac{\Delta E_h}{V} \tag{1.4}$$

where ΔE_d is the London dispersion force, ΔE_p is the dipole force, and ΔE_h is the hydrogen bonding force component. In terms of solubility parameters

$$\delta_{sp}^2 = \delta_d^2 + \delta_p^2 + \delta_h^2 \tag{1.5}$$

where δ_d, δ_p and δ_h are given as

$$\delta_d = \left(\frac{\Delta E_d}{V} \right)^{1/2} \tag{1.6}$$

$$\delta_p = \left(\frac{\Delta E_p}{V} \right)^{1/2} \tag{1.7}$$

$$\delta_h = \left(\frac{\Delta E_h}{V} \right)^{1/2} \tag{1.8}$$

and they are the dispersion force, dipole, and hydrogen bonding component of the overall solubility parameter, δ_{sp}, respectively (Hansen and Beerbower, 1971). Furthermore, $\delta_{sp}, \delta_d, \delta_p,$ and δ_h can be calculated by applying additivity rules to the structural components of the repeat unit of the macromolecules and those of solvents by the following equations (Van Krevelan, 1976).

$$\delta_{sp} = \sqrt{\frac{\sum E_{coh,i}}{V}} \tag{1.9}$$

$$\delta_d = \frac{\sum F_{di}}{V} \tag{1.10}$$

$$\delta_p = \sqrt{\frac{\sum F_{pi}^2}{V}} \tag{1.11}$$

$$\delta_h = \sqrt{\frac{\sum E_{hi}}{V}} \tag{1.12}$$

1.3 Examples of the solubility parameter

Numerical values assigned to each structural component of organic compounds are listed in Tables 1.1 and 1.2. Using these numerical values, the overall solubility parameter δ_{sp} and each of its components can be calculated when the formula of the polymer repeat unit is known.

Problem 1:

The structure of the repeat unit of cellulose acetate can be written as

$(CH_2)_4(CH)_{20}(O)_8(OH)_{2.19}(OOCCH_3)_{9.81}$ with four rings

Calculate the overall solubility parameter and its components.

Answer:

$$\sum V_i \times 10^6 = (4 \times 16.1) + (20 \times -1.0) + (8 \times 3.8) + (2.19 \times 10.0) + (9.81 \times 18.0) + (9.81 \times 33.5) = 799.02$$

$$\sum E_{cohi} = (4 \times 4937) + (20 \times 3431) + (8 \times 3347) + (2.19 \times 29{,}790) + (9.81 \times 17{,}991) + (9.81 \times 4707) = 403{,}051$$

$$\delta_{sp} = \sqrt{\frac{403051}{799.02 \times 10^{-6}}} = 22.46 \left(\frac{\text{MJ}}{\text{m}^3}\right)^{1/2}$$

$$\sum V_{gi} \times 10^6 = (4 \times 15.9) + (20 \times 9.5) + (8 \times 10.0) + (2.19 \times 9.7) + (9.81 \times 23.0) + (9.81 \times 23.9) = 814.93$$

Table 1.1 Group contributions to E_{cohi} and V_i.

Structural group	E_{cohi} J/mol	$V_i \times 10^6$ m^3/mol
$-CH_3$	4707	33.5
$-CH_2-$	4937	16.1
$>CH-$	3431	−1.0
$>C<$	1464	−19.2
$H_2C=$	4310	28.5
$-CH=$	4310	13.5
$>C=$	4310	−5.5
$HC\equiv$	3849	27.4
$-C\equiv$	7071	6.5
Phenyl	31,924	71.4
Phenylene (o,m,p)	31,924	52.4
Phenyl (trisubstituted)	31,924	33.4
Phenyl (tetrasubstituted)	31,942	14.4
Phenyl (pentasubstituted)	31,942	−4.6
Phenyl (hexasubstituted)	31,942	−23.6
Conjugation in ring for each double bond	1674	−2.2
Halogen attached to carbon atom with double bond	$0.8 \times E_{cohi}$ for halogen	4.0
$-F$	4184	18.0
$-F$ (disubstituted)	3556	20.0
$-F$ (trisubstituted)	2301	22.0
$-CF_2-$ (for perfluoro compounds)	4268	23.0
$-CF_3-$ (for perfluoro compounds)	4268	57.5
$-Cl$	11,548	24.0
$-Cl$ (disubstituted)	9623	26.0
$-Cl$ (trisubstituted)	7531	27.3
$-Br$	15,481	30.0
$-Br$ (disubstituted)	12,343	31.0
$-Br$ (trisubstituted)	10,669	32.4
$-I$	19,037	31.5
$-I$ (disubstituted)	16,736	33.5
$-I$ (trisubstituted)	16,318	37.0
$-CN$	25,522	24.0
$-OH$	27,790	10.0
$-OH$ (disubstituted or on adjacent C atoms)	21,840	13.0
$-O-$	8347	3.8
$-CHO$ (aldehyde)	21,338	22.3
$-CO-$	17,364	10.8
$-COOH$	27,614	28.5
$-CO_2-$	17,991	18.0

(Continued)

Table 1.1 (Continued)

Structural group	E_{cohi} J/mol	$V_i \times 10^6$ m^3/mol
—CO_3— (carbonate)	17,573	22.0
—C_2O_3— (anhydride)	30,543	30.0
HCOO— (formate)	17,991	32.5
—CO_2CO_2— (oxalate)	26,778	37.3
—HCO_3	12,552	18.0
—COF	13,389	29.0
—COCl	17,573	38.1
—COBr	24,142	41.6
—COI	29,288	48.7
—NH_2	12,552	19.2
—NH—	8368	4.5
—N<	4184	−9.0
—N=	11,715	5.0
—$NHNH_2$	21,966	—
—NNH_2	16,736	16
—NHNH—	16,736	16
—N_2 (diazo)	8368	23
—N=N—	4184	—
>C=N—N=C<	20,083	0
—N=C=N—	11,464	—
—NC	18,828	23.1
—NF_2	7657	33.1
—NF—	5063	24.5
—$CONH_2$	41,840	17.5
—CONH—	33,472	9.5
—CON<	29,497	−7.7
HCON<	27,614	11.3
HCONH—	43,932	27.0
—NHCOO—	26,359	18.5
—NHCONH—	50,208	—
—CONHNHCO—	46,861	19.0
—NHCON<	41,840	—
>NCON<	20,920	−14.5
NH_2COO—	36,987	—
—NCO	28,451	35.0
—ONH_2	19,037	20.0
>C=NOH	25,104	11.3
—CH=NOH	25,104	24.0
—NO_2 (aliphatic)	29,288	24.0
—NO_2 (aromatic)	15,355	32.0
—NO_3	20,920	33.5

(Continued)

Table 1.1 (Continued)

Structural group	E_{cohi} J/mol	$V_i \times 10^6$ m^3/mol
$-NO_2$ (nitrite)	11,715	33.5
$-NHNO_2$	39,478	28.7
$-NNO-$	27,196	10
$-SH$	14,435	28.0
$-S-$	14,142	12
$-S_2-$	23,849	23.0
$-S_3-$	13,389	47.2
$-SO_2-$	39,120	23.6
$>SO$	39,120	–
SO_3	18,828	27.6
SO_4	28,451	31.6
$-SO_2Cl$	37,028	43.5
$-SCN$	20,083	37.0
$-NCS$	25,104	40.0
P	9414	−1.0
PO_3	14,226	22.7
PO_4	20,920	28.0
$PO_3(OH)$	31,798	32.2
Si	3389	0
SiO_4	21,757	20.0
B	13,807	−2.0
BO_3	0	20.4
Al	13,807	−2.0
Ga	13,807	−2.0
In	13,807	−2.0
Tl	13,807	−2.0
Ge	5146	−1.5
Sn	11,297	1.5
Pb	17,154	2.5
As	12,970	7.0
Sb	16,318	8.9
Bi	21,338	9.5
Se	17,154	16.0
Te	20,083	17.4
Zn	14,477	2.5
Cd	17,782	6.5
Hg	22,803	7.5

Table 1.2 Group contribution to solubility parameter component (Van Krevelen and te Nijenhuis, 2009).

Structural group	F_{di} $(MJm^3)^{1/2}/mol$	F_{pi} $(MJm^3)^{1/2}/mol$	$F_{hi}J/mol$	$V_{gi} \times 10^6$ m^3/mol
$-CH_3$	420	0	0	23.9
$-CH_2-$	270	0	0	15.9
$>CH-$	80	0	0	9.5
$>C<$	−70	0	0	4.6
$=CH_2$	400	0	0	—
$=CH-$	200	0	0	13.1
$=C<$	70	0	0	—
Cyclohexyl	1620	0	0	90.7
Phenyl	1430	110	0	72.7
Phenylene (o,m,p)	1270	110	0	65.5
$-F$	220	—	—	10.9
$-Cl$	450	550	400	19.9
$-Br$	550	—	—	—
$-CN$	430	1100	2500	19.5
$-OH$	210	500	20,000	9.7
$-O-$	100	400	3000	10.0
$-CHO$	470	800	4500	—
$-CO-$	290	770	2000	13.4
$-COOH$	530	420	10,000	23.1
$-COO-$	390	490	7000	23.018.25 (acrylic)
$-COOH$	530	—	—	—
$-NH_2$	280	—	8400	—
$-NH-$	160	210	3100	12.5
$-N<$	20	800	5000	6.7
$-CONH-$ (aliphatic)	450	—	19,485	24.9
$-CONH-$ (aromatic) `	450	980	32,476	24.9
$-CONHNHCO-$ (aromatic)	900	—	44,472	49.8
$-NO_2$	500	1070	1500	—
$-S-$	440	—	—	17.8
$-SO_2-$	591	—	13,489	31.8
$=PO_4-$	740	1890	13,000	—
Ring	190	—	—	—
One plane of symmetry	—	0.5 ×		
	—	0.25 ×	—	—
More planes of symmetry	—	0 ×	0 ×	—

$$\sum F_{di} = (4 \times 270) + (20 \times 80) + (8 \times 100) + (2.19 \times 210) + (9.81 \times 390) + (9.81 \times 420) + (4 \times 190) = 12{,}648$$

$$\delta_d = \frac{12648}{814.93 \times 10^{-6}} = 15.52 \left(\frac{\text{MJ}}{\text{m}^3}\right)^{1/2}$$

$$\sum F_{pi}^2 = (4 \times 0)^2 + (20 \times 0)^2 + (8 \times 400)^2 + (2.19 \times 500)^2 + (9.81 \times 490)^2 + (9.81 \times 0)^2 = 34{,}545{,}313$$

$$\delta_p = \frac{\sqrt{34{,}545{,}313}}{814.93 \times 10^{-6}} = 7.21 \left(\frac{\text{MJ}}{\text{m}^3}\right)^{1/2}$$

$$\sum E_{hi} = (4 \times 0) + (20 \times 0) + (8 \times 3000) + (2.19 \times 20{,}000) + (9.81 \times 7000) + (9.81 \times 0) = 136{,}470$$

$$\delta_h = \sqrt{\frac{136{,}470}{814.93 \times 10^{-6}}} = 12.94 \left(\frac{\text{MJ}}{\text{m}^3}\right)^{1/2}$$

Problem 2:

Calculate the solubility parameters of (1) aromatic polyamide, (2) aromatic polyhydrazide, and (3) aromatic polyamidehydrazide. The structures of the polymers are

1. $NH\varphi_m NHCO\varphi_r CO$ [including 2(aromatic CONH and 2φ)] (φ = phenylene)
2. $NHNHCO\varphi_m CONHNHCO\varphi_r CO$ [including 2(aromatic CONHNHCO) and 2φ]
3. $NH\varphi_r CONHNHCO\varphi_r CO$

Answer:

1. Aromatic polyamide

$$\sum V_i \times 10^6 = (2 \times 52.4) + (2 \times 9.5) = 123.8$$

$$\sum E_{cohi} = (2 \times 31{,}924) + (2 \times 33{,}472) = 130{,}792$$

$$\delta_{sp} = \sqrt{\frac{130{,}792}{123.8 \times 10^{-6}}} = 32.50 \left(\frac{\text{MJ}}{\text{m}^3}\right)^{1/2}$$

$$\sum V_{gi} \times 10^6 = (2 \times 65.5) + (2 \times 24.9) = 180.8$$

$$\sum F_{di} = (2 \times 1270) + (2 \times 450) = 3440$$

$$\delta_d = \frac{3440}{180.8 \times 10^{-6}} = 19.03 \left(\frac{\text{MJ}}{\text{m}^3}\right)^{1/2}$$

$$\sum F_{pi}^2 = (2 \times 110)^2 + (2 \times 980)^2 = 3{,}899{,}000$$

$$\delta_p = \frac{\sqrt{3{,}899{,}000}}{180.8 \times 10^{-6}} = 10.91 \left(\frac{\text{MJ}}{\text{m}^3}\right)^{1/2}$$

$$\sum E_{hi} = (2 \times 0) + (2 \times 32{,}476) = 64{,}952$$

$$\delta_h = \sqrt{\frac{64{,}952}{180.8 \times 10^{-6}}} = 18.95 \left(\frac{\text{MJ}}{\text{m}^3}\right)^{1/2}$$

2. Aromatic polyhydrazide

$$\sum V_i \times 10^6 = (2 \times 52.4) + (2 \times 19.0) = 142.8$$

$$\sum E_{cohi} = (2 \times 31{,}924) + (2 \times 46{,}861) = 157{,}570$$

$$\delta_{sp} = \sqrt{\frac{157{,}570}{142.8 \times 10^{-6}}} = 33.22 \left(\frac{\text{MJ}}{\text{m}^3}\right)^{1/2}$$

$$\sum V_{gi} \times 10^6 = (2 \times 65.5) + (2 \times 49.8) = 230.6$$

$$\sum F_{di} = (2 \times 1270) + (2 \times 900) = 4340$$

$$\delta_d = \frac{4340}{230.6 \times 10^{-6}} = 18.82 \left(\frac{\text{MJ}}{\text{m}^3}\right)^{1/2}$$

$$\sum F_{pi}^2 = (2 \times 110)^2 + (2 \times -)^2 \ldots$$

$$\delta_p = \ldots$$

$$\sum E_{hi} = (2 \times 0) + (2 \times 44{,}472) = 88{,}944$$

$$\delta_h = \sqrt{\frac{88{,}944}{230.6 \times 10^{-6}}} = 19.64 \left(\frac{\text{MJ}}{\text{m}^3}\right)^{1/2}$$

3. Aromatic polyamidehydrazide

$$\delta_{sp} = \frac{32.50 + 33.22}{2} = 32.86\left(\frac{\mathrm{MJ}}{\mathrm{m}^3}\right)^{1/2}$$

$$\delta_d = \frac{19.03 + 18.82}{2} = 18.93\left(\frac{\mathrm{MJ}}{\mathrm{m}^3}\right)^{1/2}$$

$$\delta_p = \cdots$$

$$\delta_h = \frac{18.95 + 19.64}{2} = 19.30\left(\frac{\mathrm{MJ}}{\mathrm{m}^3}\right)^{1/2}$$

Problem 3:

The structure of the repeat unit of polyethylene terephthalate is $\varphi_p COOCH_2CH_2COO$ [including $1(\varphi)$, $2(COO)$, and $2(CH_2)$] Calculate the overall solubility parameter and its components.

Answer:

$$\sum V_i \times 10^6 = (1\times 52.4) + (2\times 18.0) + (2\times 16.1) = 120.6$$

$$\sum E_{cohi} = (1\times 31{,}924) + (2\times 17{,}991) + (2\times 4937) = 77{,}780$$

$$\delta_{sp} = \sqrt{\frac{77{,}780}{120.6\times 10^{-6}}} = 25.40$$

$$\sum V_{gi} \times 10^6 = (1\times 65.5) + (2\times 23.0) + (2\times 15.9) = 143.3$$

$$\sum F_{di} = (1\times 1270) + (2\times 390) + (2\times 270) = 2590$$

$$\delta_d = \frac{2590}{143.3\times 10^{-6}} = 18.07\left(\frac{\mathrm{MJ}}{\mathrm{m}^3}\right)^{1/2}$$

$$\sum F_{pi}^2 = (1\times 110)^2 + (2\times 490)^2 + (2\times 0)^2 = 972{,}500$$

$$\delta_p = \frac{\sqrt{972{,}500}}{143.3\times 10^{-6}} = 6.88\left(\frac{\mathrm{MJ}}{\mathrm{m}^3}\right)^{1/2}$$

$$\sum E_{hi} = (1\times 0) + (2\times 7000) + (2\times 0) = 14{,}000$$

$$\delta_h = \sqrt{\frac{14{,}000}{143.3\times 10^{-6}}} = 9.88\left(\frac{\mathrm{MJ}}{\mathrm{m}^3}\right)^{1/2}$$

Problem 4:

Calculate the overall solubility parameters of (1) polyvinyl alcohol and (2) polyvinyl acetal and their components.

Answer:

1. Polyvinyl alcohol

The repeat unit of polyvinyl alcohol includes 1 (CH_2), 1 (CH), and 1 (OH). Therefore

$$\sum V_i \times 10^6 = (1 \times 16.1) + (1 \times -1.0) + (1 \times 10.0) = 25.1$$

$$\sum E_{cohi} = (1 \times 4937) + (1 \times 3431) + (1 \times 27{,}790) = 36{,}158$$

$$\delta_{sp} = \sqrt{\frac{36{,}158}{25.1 \times 10^{-6}}} = 37.95 \left(\frac{\text{MJ}}{\text{m}^3}\right)^{1/2}$$

$$\sum V_{gi} \times 10^6 = (1 \times 15.9) + (1 \times 9.5) + (1 \times 9.7) = 35.1$$

$$\sum F_{di} = (1 \times 270) + (1 \times 80) + (1 \times 210) = 560$$

$$\delta_d = \frac{560}{35.1 \times 10^{-6}} = 16.0 \left(\frac{\text{MJ}}{\text{m}^3}\right)^{1/2}$$

$$\sum F_{pi}^2 = (1 \times 0)^2 + (1 \times 0)^2 + (1 \times 500)^2 = 250{,}000$$

$$\delta_p = \frac{\sqrt{250{,}000}}{35.1 \times 10^{-6}} = 14.3 \left(\frac{\text{MJ}}{\text{m}^3}\right)^{1/2}$$

$$\sum E_{hi} = (1 \times 0) + (1 \times 0) + (1 \times 20{,}000) = 20{,}000$$

$$\delta_h = \sqrt{\frac{20{,}000}{35.1 \times 10^{-6}}} = 23.9 \left(\frac{\text{MJ}}{\text{m}^3}\right)^{1/2}$$

2. Polyvinyl acetal

The repeat unit of polyvinyl acetal includes 1 (CH_3), 2 (CH_2), 3 (CH), 2 (O), and 1 ring. Therefore

$$\sum V_i \times 10^6 = (1 \times 33.5) + (2 \times 16.1) + (3 \times -1.0) + (2 \times 3.8) = 70.3$$

$$\sum E_{cohi} = (1 \times 4707) + (2 \times 4937) + (3 \times 3431) + (2 \times 8347) = 31{,}568$$

$$\delta_{sp} = \sqrt{\frac{31{,}568}{70.3 \times 10^{-6}}} = 21.19 \left(\frac{\text{MJ}}{\text{m}^3}\right)^{\frac{1}{2}}$$

$$\sum V_{gi} \times 10^6 = (1 \times 23.9) + (2 \times 15.9) + (3 \times 9.5) + (2 \times 10.0) = 104.2$$

$$\sum F_{di} = (1 \times 420) + (2 \times 270) + (3 \times 80) + (2 \times 100) + (1 \times 190) = 1590$$

$$\delta_d = \frac{1590}{104.2 \times 10^{-6}} = 15.26 \left(\frac{\text{MJ}}{\text{m}^3}\right)^{1/2}$$

$$\sum F_{pi}^2 = (1 \times 0)^2 + (2 \times 0)^2 + (3 \times 0)^2 + (2 \times 400)^2 = 640{,}000$$

$$\delta_p = \frac{\sqrt{640{,}000}}{104.2 \times 10^{-6}} = 7.68 \left(\frac{\text{MJ}}{\text{m}^3}\right)^{1/2}$$

$$\sum E_{hi} = (1 \times 0) + (2 \times 0) + (3 \times 0) + (2 \times 3000) = 6000$$

$$\delta_h = \sqrt{\frac{6000}{104.2 \times 10^{-6}}} = 7.59 \left(\frac{\text{MJ}}{\text{m}^3}\right)^{1/2}$$

Problem 5:

Calculate the overall solubility parameter of polydimethyl siloxane.

Answer:

The repeat unit includes 2 (CH_3), 1 (O), and 1 (Si). Therefore

$$\sum V_i \times 10^6 = (2 \times 33.5) + (1 \times 3.8) + (1 \times 0) = 70.7$$

$$\sum E_{cohi} = (2 \times 4707) + (1 \times 8347) + (1 \times 3389) = 21{,}150$$

$$\delta_{sp} = \sqrt{\frac{21{,}150}{70.7 \times 10^{-6}}} = 17.30 \left(\frac{\text{MJ}}{\text{m}^3}\right)^{\frac{1}{2}}$$

Problem 6:

PVA and PDMS membranes are considered for pervaporation membrane to separate ethanol and water. Which one (ethanol or water) permeates the membrane preferentially?

Answer:

PVA membrane

$$|\delta_{sp,water} - \delta_{sp,PVA}| = 10.05$$

$$|\delta_{sp,ethanol} - \delta_{sp,PVA}| = 11.75$$

Interaction of PVA with water is stronger than with ethanol. Hence, water preferentially permeates through the membrane.

PDMS membrane

$$|\delta_{sp,water} - \delta_{sp,PDMS}| = 31.3$$

$$|\delta_{sp,ethanol} - \delta_{sp,PDMS}| = 8.9$$

Interaction of PDMS with ethanol is stronger than with water. Hence, ethanol preferentially permeates through the PDMS membrane.

Solubility parameters for some chosen solvents and polymers are shown in Tables 1.3 and 1.4, respectively.

Table 1.3 Solubility parameter of solvent (Solubility Parameters: Theory and Application, 1984).

Solvent	Solubility parameter, δ_{sp}, $(MJ/m^3)^{1/2}$
n-Pentane	14.4
n-Hexane	14.9
n-Heptane	15.3
Diethylether	15.4
1,1,1-Trichloroethane	15.8
n-Dodecane	16.0
White spirit	16.1
Turpentine	16.6
Cyclohexane	16.8
Amyl acetate	17.1
Carbon tetrachloride	18.0
Xylene	18.2
Ethyl acetate	18.2
Toluene	18.3
Tetrahydrofuran	18.5
Benzene	18.7
Chloroform	18.7
Trichloroethylene	18.7
Cellosolve acetate	19.1
Methyl ethyl ketone	19.3
Acetone	19.7
Diacetone alcohol	20.0
Ethylene dichloride	20.2
Methylene chloride	20.2
Butyl cellosolve	20.2

(Continued)

Table 1.3 (Continued)

Solvent	Solubility parameter, δ_{sp}, $(MJ/m^3)^{1/2}$
Pyridine	21.7
Cellosolve	21.9
Morpholine	22.1
Dimethylformamide	24.7
n-Propyl alcohol	24.9
Ethyl alcohol	26.2
Dimethyl sulfoxide	26.4
n-Butyl alcohol	28.7
Methyl alcohol	29.7
Propylene glycol	30.7
Ethylene glycol	34.9
Glycerol	36.2
Water	48.0

Table 1.4 Solubility parameter of polymers in $(MJ/m^3)^{1/2}$.

Polymers	Structural components	δ_{sp}	δ_d	δ_p	δ_h
Cellulose acetate	$(CH_2)_4(CH)_{20}(O)_8(OH)_{2.19}(OOCCH_3)_{9.81}$with four rings	22.46	15.52	7.21	12.94
Cellulose triacetate	$(CH_2)_4(CH)_{20}(O)_8(OH)(OOCCH_3)_{11}$with four rings	24.61	15.55	6.84	11.87
Cellulose nitrate	$(CH_2)_4(CH)_{20}(O)_8(NO_3)_{12}$with four rings	27.71	–	–	–
Cellulose	$(CH_2)_4(CH)_{20}(O)_8(OH)_{12}$with four rings	49.26	15.02	15.11	24.22
Nylon 6	$(CH_2)_5$(CONH, aliphatic)	25.42	17.24	–	13.66
Polyamide	$NH\varphi_mNHCO\varphi_rCO$	32.50	19.03	10.91	18.95
Carboxylated polyamide	$(NH\varphi_mNHCO\varphi_{m,p}(COOH)CO)_{1.3}(NH\varphi_mNHCO\varphi_pCO)_{0.7}$	33.73	19.15	–	19.27
Polyhydrazide	$NHNHCO\varphi_mCONHNHCO\varphi_rCO$	33.22	18.82	–	19.64
Copolyamide hydrazide	$NH\varphi_rCONHNHCO\varphi_rCO$	32.86	18.93	–	19.30
Polyalanine	$(CH_3)(CH)(CONH)$	31.48	16.30	–	18.28
Polyacryilic acid	$(CH_2)(CH)(COOH)$	28.73	18.14	8.66	14.46
Polymetacrylic acid	$(CH_3)(CH_2)(C)(COOH)$	25.64	17.04	6.22	12.17
Polyvinyl alcohol	$(CH_2)(CH)(OH)$	37.95	16.0	14.3	23.9
Polyvinyl acetate	$(CH_3)(CH_2)(CH)(COO)$	21.60	16.04	6.78	9.84
Polyvinyl formal	$(CH_2)_3(CH)_2(O)_2$with one ring	22.94	15.69	9.23	8.32
Polyvinyl acetal	$(CH_3)(CH_2)_2(CH)_3(O)_2$with one ring	21.19	15.26	7.68	7.59
Polyvinyl butyral	$(CH_3)(CH_2)_4(CH)_3(O)_2$with one ring	20.11	15.66	5.88	6.64
Polyethylene	$(CH_2)_2$	17.51	16.98	0	0
Polypropylene	$(CH_3)(CH_2)(CH)$	16.40	15.62	0	0
Polystyrene	$(CH_2)(CH)\ \varphi$	21.58	18.14	0	0

(Continued)

Table 1.4 (Continued)

Polymers	Structural components	δ_{sp}	δ_d	δ_p	δ_h
Polyvinyl chloride	$(CH_2)(CH)Cl$	22.57	17.66	12.14	2.97
Polymethyl methacrylate	$(CH_3)_2(CH_2)(C)(COO)$	20.32	16.52	5.66	8.99
Polyethyl acrylate	$(CH_3)(CH_2)_2(CH)(COO)$	20.86	17.14	5.87	9.16
Poly(2-hydroxyethylmethacrylate)	$(CH_3)(CH_2)_3(C)(COO)(OH)$	27.15	16.90	6.72	16.10
Polyethylene glycol	$(CH_2)_2O$	19.16	15.31	9.57	8.47
Polypropylene glycol	$(CH_3)(CH_2)(CH)O$	17.70	14.67	6.75	7.11
Polyacrylonitrile	$(CH_2)(CH)(CN)$	29.44	17.37	24.50	7.46
Polybutadiene	$(CH_2)_2({=}CH-)_2$	17.67	16.2	0	0
Polyisobutylene	$(CH_3)_2(CH_2)(C)$	15.73	15.23	0	0
Polyacetylene	$({=}C-)_2$	17.87	15.27	0	0
Polyvinylidene chloride	$(CH_2)(C)(Cl)_2$	22.90	18.24	18.24	3.64
Polychlorotrifluoroethylene	(C)(CF_2 perfluoro compounds)(Cl)(F)	21.65	15.70	–	–
Polytetrafluoroethylene	(CF_2 perfluoro compounds)$_2$	13.62	14.02	–	–
Polyvinylidene fluoride	(CH_2)(CF_2 perfluoro compounds)	15.34	15.13	–	–
Polyethyleneterephthalate	$(CH_2)_2\varphi m(COO)_2$	25.40	18.07	6.88	6.99
Polybutyleneterephthalate	(CH_2)$_4$(phenylene)(COO)$_2$	23.95	17.88	5.63	6.32
Polycarbonate	$(CH_3)_2(C)(\varphi m)_2(COO)(O)$	23.47	17.56	2.92	6.80
Polyethylacrylate	$(CH_3)(CH_2)_2(CH)(COO)$	20.86	17.14	5.87	9.16
Polysulfone	$(CH_3)_2(C)\ (\varphi m)_3(O)_2(SO_2)$	25.41	17.86	–	8.05
Polyethersulfone	$(\varphi m)_2(O)(SO_2)$	28.36	18.70	–	9.77
Polyimide P84	(CH_3)$_{0.8}$(CH_2)$_{0.2}$(phenylene)$_{0.4}$(phenyl, trisubstituted)$_{2.4}$($>$N$-$)$_{1.6}$(CO)$_{4.0}$	32.63	–	–	–
Polyetherimide	(CH3)$_2$(C)(phenylene)$_3$(phenyl, trisubstituted)$_2$(O)$_2$($>$N$-$)$_2$(CO)$_4$	28.93	–	–	–
Polyphenylene sulfide	(phenylene)(S)	26.75	20.53	–	–
Polyetheretherketone	(phenylene)$_3$(O)$_2$(CO)	26.12	18.70	5.04	5.90
Polyphenylene oxide	(CH_3)$_2$(phenyl trisubstituted)(O)	20.71	–	–	–

Nomenclature

$E_{coh,i}$	Structural group contribution to δ_{sp} (J/mol)
$E_{h,i}$	Structural group contribution to δ_h (J/mol)
$F_{d,i}$	Structural group contribution to δ_p $((MJm^3)^{1/2}/mol)$
$E_{p,i}$	Structural group contribution to δ_h $((MJm^3)^{1/2}/mol)$
T	Temperature (K)
V	Molar volume (m^3/mol)
V_M	Total molar volume of mixture (m^3/mol)

Greek letters

δ_{sp}	Overall solubility parameter $((MJ/m^3)^{1/2})$
δ_d	Solubility parameter, London dispersion force component $((MJ/m^3)^{1/2})$

δ_p Solubility parameter, dipole force component ($(MJ/m^3)^{1/2}$)

δ_h Solubility parameter, hydrogen bonding force component ($(MJ/m^3)^{1/2}$)

ΔE Heat of vaporization (J/mol)

ΔE_d Heat of vaporization, London dispersion force component (J/mol)

ΔE_p Heat of vaporization, dipole force component (J/mol)

ΔE_h Heat of vaporization, hydrogen bonding force component (J/mol)

ΔG Free energy of mixing (J/mol)

ΔG_M Total heat of mixing (J/mol)

ΔH Heat of mixing (J/mol)

ΔS Entropy of mixing (J/mol K)

φ Volume fraction (–)

Subscript

1,2 Components 1 and 2 of a mixture

References

Hansen, C.M., Beerbower, A., 1971. Encyclopedia of Chemical Technology, Supplement Volume. John Wiley and Sons, New York, pp. 889–910.

Hildebrand, J.H., Scott, R.L., 1950. The Solubility of Non-electrolyte, third ed. Reinhold, New York, pp. 123–124.

Small, P.A., 1953. Some factors affecting the solubility of polymers. J. Appl. Chem 3, 71–80.

Solubility Parameters: Theory and Application, 1984. J. Burke Oakland Museum of California. <https://cool.culturalheritage.org/byauth/burke/solpar/solpar2.html>.

Van Krevelan, D.W., 1976. Properties of Polymers. Elsevier, Amsterdam.

Van Krevelen, D.W., te Nijenhuis, K., 2009. Properties of Polymers, fourth ed. Elsevier, Amsterdam, p. 215.

2

Triangular phase diagram

2.1 Principles

2.1.1 Thermodynamics of polymer solution

The phase-inversion technique developed by Loeb-Sourirajan is by far the most popular method to prepare polymeric membranes with asymmetric structure. Hence it is meaningful to discuss the formation mechanism of asymmetric structure during the phase inversion process. In this process, the casting dope that is a homogeneous polymeric solution is cast into a film. In the dry phase inversion process, solvent gradually evaporates from the film and the solution composition crosses the phase boundary line, where the solution is split into two phases—the polymer-rich and polymer-lean phases. The polymer-rich phase, after further evaporation of solvent, becomes the polymer matrix of the membrane and the polymer-lean phase turns into the pores of the membrane.

In the wet phase inversion process, the cast film is immersed into a gelation media that is nonsolvent for the polymer. While the film is in the gelation media, solvent–nonsolvent exchange takes place and the polymer composition crosses the phase boundary line and the solution is split into the two, polymer-rich and polymer-lean, phases. Hereafter, the formation of the porous structure is the same as in the dry phase inversion process. The dry and wet phase inversion process may also be combined in the dry–wet phase inversion process. Fig. 2.1 shows the composition path that takes place during the phase inversion process, where α is the composition of the casting solution, β is the composition at which phase separation takes place, γ is the composition at which the polymeric film solidifies, and δ is the composition of the membrane when finally fabricated (Matsuura, 1994).

Thus, it is very important to know the composition change of the polymer solution from the cast film to the finally obtained membrane.

Since the separation of polymer solution into two phases is a thermodynamic process, the phase boundary line can be drawn by considering the thermodynamic equilibrium.

Membrane Separation Processes. DOI: https://doi.org/10.1016/B978-0-12-819626-7.00005-3

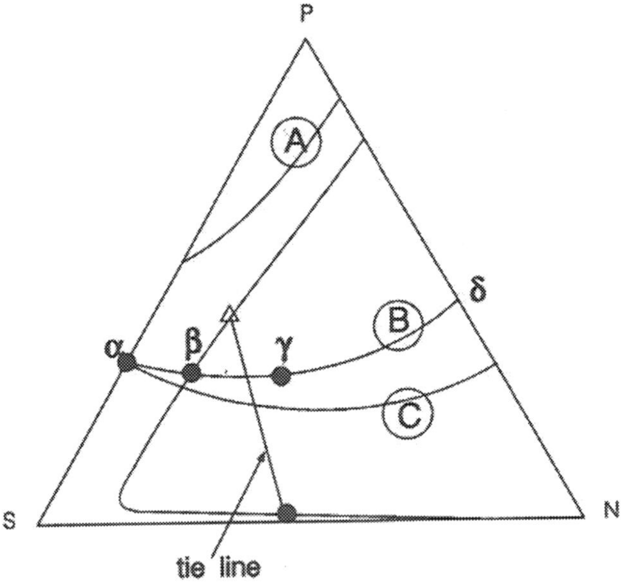

Figure 2.1 Line α–β–γ–δ is a schematic presentation of the composition path during the phase inversion process (Matsuura, 1994).

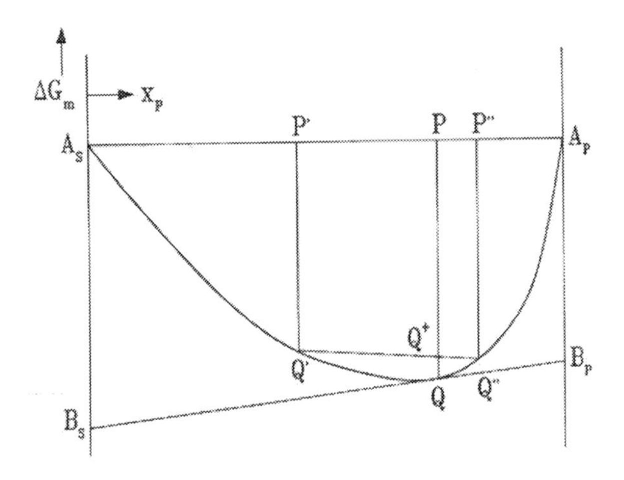

Figure 2.2 Free energy of mixing versus mole fraction of polymer I (Matsuura, 1994).

The thermodynamics of a polymer solution was discussed thoroughly by Tompa in his book published in 1956 (Tompa, 1956), based on the Flory–Higgins equation for the free energy of mixing.

Suppose the free energy of mixing, ΔG_m, for a binary mixture of solvent and polymer is given as a function of the mole fraction of polymer X_p, as drawn in Fig. 2.2. In the figure the free energy of mixing is Q, corresponding to the composition P. If this solution is separated into two phases, whose compositions are P′ and P″ (the free energies of mixing of these two phases are

Q′ and Q″, respectively), the overall free energy of the separate phases is given by Q$^+$ by the lever rule, and Q$^+$ is higher than Q as far as the ΔG_m curve is concave upward. This means that the homogeneous polymer solution with a composition of P is thermodynamically stable and phase separation does not occur.

Furthermore,

$$\Delta G_m = x_s \Delta \mu_s + x_p \Delta \mu_p \tag{2.1}$$

where x_s and x_p are the mole fractions of solvent and polymer, respectively, at P ($x_s + x_p = 1$) and $\Delta \mu_s$ and $\Delta \mu_p$ are, respectively, chemical potential of solvent and polymer at point P.

Since

$$\left(\frac{\partial \Delta G_m}{\partial x_p}\right)_{P,T} = \Delta \mu_p - \Delta \mu_s \tag{2.2}$$

From the above two equations

$$\Delta \mu_s = \Delta G_m - x_p \left(\frac{\partial \Delta G_m}{\partial x_p}\right)_{P,T} \tag{2.3}$$

$$\Delta \mu_p = \Delta G_m + x_s \left(\frac{\partial \Delta G_m}{\partial x_p}\right)_{P,T} \tag{2.4}$$

Eqs. (2.3) and (2.4) mean that the intercepts of the tangent of ΔG_m line at $x_p = 0$ and 1 ($A_s B_s$ and $A_p B_p$), respectively, correspond to $\Delta \mu_s$ and $\Delta \mu_p$.

When the free energy of mixing is given by a curve as illustrated in Fig. 2.3, and the mixture P is very close to P′ (or P″) the solution is stable since the curve is concave upward. But when P is away from P′ (or P″), P goes into a region where the free energy curve becomes concave downward and Q$^+$ becomes lower than Q. Then the solution P tends to separate into two phases whose compositions are P′ and P″. Note that *P* moves from concave upward to concave downward region at the inflection point of the curve. Hence the region from P′ to the first inflection point is called the meta stable region, from the first to the second inflection point it is called unstable region, and the second inflection point to P″ is again called the meta stable region.

Furthermore, since the intercepts of tangent at $x_p = 0$ and 1 are the same for the two coexisting phases P′ and P″,

$$\Delta \mu_s' = \Delta \mu_s'' \tag{2.5}$$

and

$$\Delta \mu_p' = \Delta \mu_p'' \tag{2.6}$$

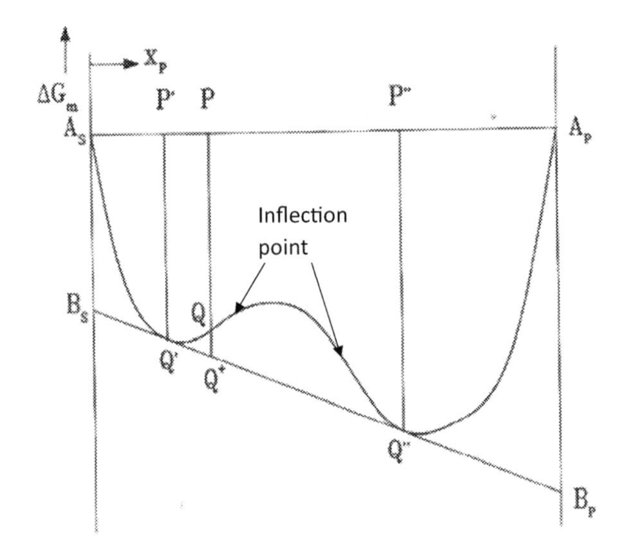

Figure 2.3 Free energy of mixing versus mole fraction of polymer II (Matsuura, 1994).

The boundary between metastable and unstable regions is given from the second derivative of ΔG_m equal to zero. Then differentiating Eq. (2.2) once more,

$$\frac{\partial \Delta \mu_s}{\partial x_p} = \frac{\partial \Delta \mu_p}{\partial x_p} \tag{2.7}$$

In conjunction with the Gibbs–Duhem relation,

$$x_s d\mu_s + x_p d\mu_p = 0 \tag{2.8}$$

$$\frac{\partial \Delta \mu_s}{\partial x_p} = \frac{\partial \Delta \mu_p}{\partial x_p} = 0 \tag{2.9}$$

When the temperature is increased, the two minima appearing on the free energy curve at Q′ and Q″ get closer and eventually meet with each other. At this point, the third derivative of ΔG_m becomes zero. Again in conjunction with the Gibbs–Duhem relation,

$$\frac{\partial^2 \Delta \mu_s}{\partial x_p^2} = \frac{\partial^2 \Delta \mu_p}{\partial x_p^2} = 0 \tag{2.10}$$

Such a point is called the plait point.

Using the equations derived above the phase diagram of the binary system can be drawn.

At a given temperature, two mole fractions x_p' and x_p'' that can satisfy Eqs. (2.5) and (2.6) are searched for. Then, a similar search is made for another temperature and this process is continued. By plotting the mole fractions so obtained versus temperature, a solid line on Fig. 2.4 can be drawn. Next, mole fractions are searched for to satisfy Eq. (2.9), and by plotting the mole fraction so obtained versus temperature, a broken line on Fig. 2.4 can be drawn. Finally, the solid and broken lines merge at the plait point, where Eq. (2.10) is satisfied.

Fig. 2.4 is the phase diagram for a binary system and its meaning is as follows. At a given temperature T, the binary mixture is homogeneous and stable in the range from the A_s axis to the solid line (or $x_p = 0$ to $x_p = x_p'$). The mixture from the solid line to the broken line is metastable. The mixture remains against the spontaneous separation into two phases. When the local solution composition crosses the broken line by perturbation and goes into the region between the two broken lines, the solution becomes unstable and tends to separate spontaneously into two phases whose compositions are on the solid line (x_p' and x_p'' in Fig. 2.4). As the composition moves toward the right, it goes into another metastable region between the broken and solid lines and becomes homogeneous and stable again from a solid line to the A_p axis (from $x_p = x_p''$ to $x_p = 1$). With an increase in T, the homogeneous and stable region becomes broader, and at $T = T_c$ a plait point appears, where the solid and broken lines merge at $x_p = x_{pc}$. At a temperature above T_c the mixture is homogeneous and stable in the entire range from $x_p = 0$ to 1.

As for a ternary system including nonsolvent (n), solvent (s), and polymer (p), the following equations can be used.

Corresponding to Eqs. (2.5) and (2.6) of the binary system to draw the solid line,

$$\Delta\mu_n' = \Delta\mu_n'' \tag{2.11a}$$

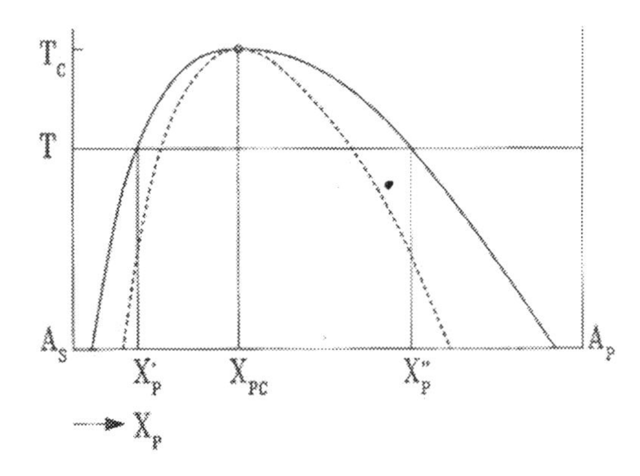

Figure 2.4 Phase diagram of a binary mixture (Matsuura, 1994).

$$\Delta\mu_s' = \Delta\mu_s'' \tag{2.11b}$$

$$\Delta\mu_p' = \Delta\mu_p'' \tag{2.11c}$$

Corresponding to Eq. (2.9) of the binary system to draw the broken line,

$$G_{ss}G_{pp} = G_{sp}^2 \tag{2.12}$$

And corresponding to Eq. (2.10) to obtain the plait point,

$$G_{sss} - 3gG_{ssp} + 3g^2G_{spp} - g^3G_{ppp} = 0 \tag{2.13}$$

where G_{ij} is the partial derivative of ΔG_m with respect to x_i and x_j. G_{ijk} is the partial derivative of ΔG_m with respect to x_i, x_j and x_k. g is G_{ss}/G_{sp}.

A typical example of the solutions to Eqs. (2.11)–(2.13) is illustrated in Fig. 2.5. Eqs. (2.11a)–(2.11c) produce a solid line in Fig. 2.5. This line is called the binodial line. A pair of compositions is obtained as those of coexisting phases, and the straight line connecting these compositions is called the tie line. The solution to Eq. (2.12) produces a broken line in Fig. 2.5. This line is called the spinodial line. The binodial and spinodial lines merge at the plait point whose composition is obtained by solving Eq. (2.13). The meaning of the binodial and spinodial lines and the tie line becomes clear in analogy to the binary system shown in Fig. 2.4. A ternary mixture whose composition lies in the region surrounded by the NS and SP axes and the solid line is homogeneous and stable. A mixture whose composition is in the region surrounded by the solid (binodial) and broken (spinodial) lines is metastable. A mixture whose composition is inside

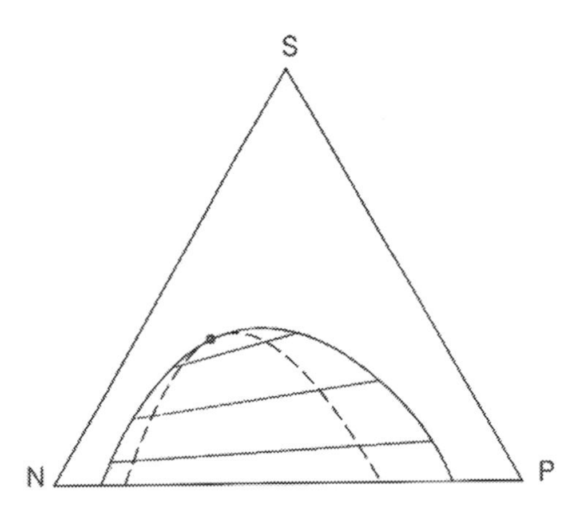

Figure 2.5 Phase diagram of a ternary system (Matsuura, 1994).

the region surrounded by the broken (spinodial) line is unstable and tends to separate spontaneously into two phases, whose compositions are at the ends of the tie line.

2.1.2 Solutions for the ternary system using Flory–Huggins equations

The solutions for the ternary system are now attempted using the Flory–Huggins equations for the free energy of mixing.

$$\frac{\Delta G_m}{RT} = n_n \ln\varphi_n + n_s\varphi_s + n_p\varphi_p + (\chi_{ns}\varphi_n\varphi_s + \chi_{sp}\varphi_s\varphi_p + \chi_{pn}\varphi_p\varphi_n) \times (m_n n_n + m_s n_s + m_p n_p) \tag{2.14}$$

where n_i is the number of moles of the component i, φ_i is the volume fraction of the component i, m_i is the ratio of the molar volume of the component i to that of the solvent (in other words $m_s = 1$), and χ_{ij} is the interaction constant between the components i and j. Rearranging Eq. (2.14), the free energy can be written as

$$\frac{\Delta G_m}{RT} = \left\{\sum \frac{\varphi_i}{m_i}\ln\varphi_i + \sum \chi_{ij}\varphi_i\varphi_j\right\} \times \sum m_i n_i \tag{2.15}$$

where the summation over i and j is to be taken over all different pairs of i and j.

Since

$$\frac{\Delta\mu_n}{RT} = \frac{\partial \Delta G_m/RT}{\partial n_n} \tag{2.16}$$

$$\frac{\Delta\mu_n}{RT} = \ln\varphi_n + 1 - m_n\sum\frac{\varphi_i}{m_i} + m_n\sum\chi_{ni}\varphi_i - m_n\sum\chi_{ij}\varphi_i\varphi_j \tag{2.17}$$

Further rearrangement yields

$$\frac{\Delta\mu_n}{RT} = \ln\varphi_n + \left(1 - \frac{m_n}{m_s}\right)\varphi_s + \left(1 - \frac{m_n}{m_p}\right)\varphi_p + m_n\left\{\chi_n\left(\varphi_s + \varphi_p\right)^2 + \chi_s\varphi_s^2\chi_p\varphi_p^2\right\} \tag{2.18}$$

Similarly,

$$\frac{\Delta\mu_s}{RT} = \ln\varphi_s + \left(1 - \frac{m_s}{m_p}\right)\varphi_p + \left(1 - \frac{m_s}{m_n}\right)\varphi_n + m_s\left\{\chi_s\left(\varphi_p + \varphi_n\right)^2 + \chi_p\varphi_p^2\chi_n\varphi_n^2\right\} \tag{2.19}$$

$$\frac{\Delta\mu_p}{RT} = \ln\varphi_p + \left(1 - \frac{m_p}{m_n}\right)\varphi_n + \left(1 - \frac{m_p}{m_s}\right)\varphi_s + m_p\left\{\chi_p(\varphi_n + \varphi_s)^2 + \chi_n\varphi_n^2\chi_s\varphi_s^2\right\} \tag{2.20}$$

where χ_n, χ_s and χ_p are given as

$$2\chi_n = \chi_{ns} + \chi_{pn} - \chi_{sp} \tag{2.21}$$

$$2\chi_s = \chi_{sp} + \chi_{ns} - \chi_{pn} \tag{2.22}$$

$$2\chi_p = \chi_{pn} + \chi_{sp} - \chi_{ns} \tag{2.23}$$

The equations for the spinodial and the plait point are, respectively,

$$\sum m_i\varphi_i - 2\sum m_i m_j(\chi_i + \chi_j)\varphi_i\varphi_j + 4m_n m_s m_p \sum \chi_i\chi_j\varphi_n\varphi_s\varphi_p = 0 \tag{2.24}$$

and

$$\sum \frac{m_i^2\varphi_i}{(1 - 2\chi_i m_i\varphi_i)^3} = 0 \tag{2.25}$$

For the coexisting two phases equations, Eq. (2.11) and the following two equations should be satisfied.

$$\varphi_n' + \varphi_s' + \varphi_p' = 1 \tag{2.26}$$

$$\varphi_n'' + \varphi_s'' + \varphi_p'' = 1 \tag{2.27}$$

Considering Eqs. (2.11), (2.26), and (2.27) there are five equations and six unknowns (φ_n', φ_s', φ_p', φ_n'', φ_s'' and φ_p''). Therefore, for a given φ_n', the other five unknowns can be found. The straight line connecting two compositions (φ_n', φ_s', φ_p') and (φ_n'', φ_s'' and φ_p'') is the tie line. The binodial line can be drawn by changing the value of φ_n'.

As for the spinodial line, there are two equations, Eq. (2.24) and

$$\varphi_n + \varphi_s + \varphi_p = 1 \tag{2.28}$$

while there are three unknown, φ_n, φ_s and φ_p. Therefore a dotted line can be drawn by changing φ_n.

A point exists on this curve that satisfies Eq. (2.25) and this is the plait point.

Problem 1:

Draw a binodial line when

$$m_n = m_s = 1 \text{ and } m_p = 100$$

and

$$\chi_{ns} = \chi_{pn} = \chi = 1.5 \text{ and } \chi_{sp} = 0$$

This means the polymer (P)'s molar volume is much larger than the nonsolvent (N) and solvent (S). The N/S and P/N interactions are weak, while the S/P interaction is strong. (Note that the interaction becomes stronger as the interaction constant (χ) decreases.)

For $\chi_{ns} = \chi_{pn} = \chi$ and $\chi_{sp} = 0$, Tompa derived the following simplified equations for the chemical potential of the three components.

$$\frac{\Delta\mu_n}{RT} = \ln\varphi_n + \left(1 - \frac{1}{m_p}\right)\varphi_p + \chi(1-\varphi_n)^2 \tag{2.29}$$

$$\frac{\Delta\mu_s}{RT} = \ln\varphi_s + \left(1 - \frac{1}{m_p}\right)\varphi_p + \chi\varphi_n^2 \tag{2.30}$$

$$\frac{\Delta\mu_p}{RT} = \ln\varphi_p - (m_p - 1)\left(1 - \varphi_p\right) + m_p\chi\varphi_n^2 \tag{2.31}$$

Answer:

Using the given numerical values,

$$\begin{aligned} &\ln\varphi_n' + \left(1 - \frac{1}{100}\right)\varphi_p' + 1.5(1-\varphi_n')^2 \\ &= \ln\varphi_n'' + \left(1 - \frac{1}{100}\right)\varphi_p'' + 1.5(1-\varphi_n'')^2 \end{aligned} \tag{2.32}$$

$$\ln\varphi_s' + \left(1 - \frac{1}{100}\right)\varphi_p' + 1.5(\varphi_n')^2 = \ln\varphi_n'' + \left(1 - \frac{1}{100}\right)\varphi_p'' + 1.5(\varphi_n'')^2 \tag{2.33}$$

$$\begin{aligned} &\ln\varphi_p' - (100 - 1)\left(1 - \varphi_p'\right) + 150(\varphi_n')^2 \\ &= \ln\varphi_p'' - (100 - 1)\left(1 - \varphi_p''\right) + 150(\varphi_n'')^2 \end{aligned} \tag{2.34}$$

Furthermore,

$$\varphi_n' + \varphi_s' + \varphi_p' = 1 \tag{2.26}$$

$$\varphi_n'' + \varphi_s'' + \varphi_p'' = 1 \tag{2.27}$$

Rearranging Eq. (2.33)

$$\ln\frac{\varphi_s''}{\varphi_s'} = -\left(1-\frac{1}{100}\right)\left(\varphi_p''-\varphi_p'\right) - 1.5\left(\left(\varphi_n''\right)^2 - \left(\varphi_n'\right)^2\right) \tag{2.35}$$

Therefore

$$-\ln\frac{\varphi_s'}{\varphi_s''} = 0.99(\varphi_p' - \varphi_p'') + 1.5\left(\left(\varphi_n'\right)^2 - \left(\varphi_n''\right)^2\right) \tag{2.36}$$

From Eq. (2.34)

$$\begin{aligned}&\frac{1}{100}\ln\varphi_p' + \left(1-\frac{1}{100}\right)\varphi_p' + 1.5\left(\varphi_n'\right)^2\\ &= \frac{1}{100}\ln\varphi_p'' + \left(1-\frac{1}{100}\right)\varphi_p'' + 1.5\left(\varphi_n''\right)^2\end{aligned} \tag{2.37}$$

Eq (2.37)−eq. (2.33) yields

$$\frac{1}{100}\ln\varphi_p' - \varphi_s' = \frac{1}{100}\ln\varphi_p'' - \varphi_s'' \tag{2.38}$$

Therefore

$$-\ln\frac{\varphi_s'}{\varphi_s''} = -\frac{1}{100}\ln\frac{\varphi_p'}{\varphi_p''} \tag{2.39}$$

Finally, Eq. (2.32)−eq. (2.33) yields

$$\begin{aligned}&\ln\varphi_n' - \varphi_s' + 1.5\left\{\left(1-\varphi_n'\right)^2 - \left(\varphi_n'\right)^2\right\}\\ &= \ln\varphi_n'' - \varphi_s'' + 1.5\left\{\left(1-\varphi_n''\right)^2 - \left(\varphi_n''\right)^2\right\}\end{aligned} \tag{2.40}$$

Rearranging,

$$-\ln\frac{\varphi_s'}{\varphi_s''} = 3\left(\varphi_n' - \varphi_n''\right) - \ln\frac{\varphi_n'}{\varphi_n''} \tag{2.41}$$

From Eqs. (2.36) and (2.41)

$$\left(\varphi_p' - \varphi_p''\right) = \frac{3\left(\varphi_n' - \varphi_n''\right) - \ln\frac{\varphi_n'}{\varphi_n''} - 1.5\left(\left(\varphi_n'\right)^2 - \left(\varphi_n''\right)^2\right)}{0.99} \tag{2.42}$$

From Eqs. (2.39) and (2.41)

$$\ln\frac{\varphi_p'}{\varphi_p''} = -100\times\left\{3\left(\varphi_n' - \varphi_n''\right) - \ln\frac{\varphi_n'}{\varphi_n''}\right\} \tag{2.43}$$

From Eq. (2.41)

$$\ln\frac{\varphi_s'}{\varphi_s''} = -\left\{3\left(\varphi_n' - \varphi_n''\right) - \ln\frac{\varphi_n'}{\varphi_n''}\right\} \tag{2.44}$$

From Eqs. (2.26), (2.27), (2.42)–(2.44), φ_s', φ_p', φ_n'', φ_s'' and φ_p'' can be obtained for a given value of φ_n' by the following steps.

1. Guess φ_n''.
2. Calculate $(\varphi_p' - \varphi_p'')$ and $\frac{\varphi_p'}{\varphi_p''}$ by Eqs. (2.42) and (2.43) and solve φ_p' and φ_p''.
3. Obtain φ_s' and φ_s'' from Eqs. (2.26) and (2.27) and calculate $\ln\frac{\varphi_s'}{\varphi_s''}$.
4. Calculate $\ln\frac{\varphi_s'}{\varphi_s''}$ from Eq. (2.44).
5. Repeat the process until $\ln\frac{\varphi_s'}{\varphi_s''}$ values calculated by steps (3) and (4) agree.

The calculation algorithm is given in Scheme 2.1.

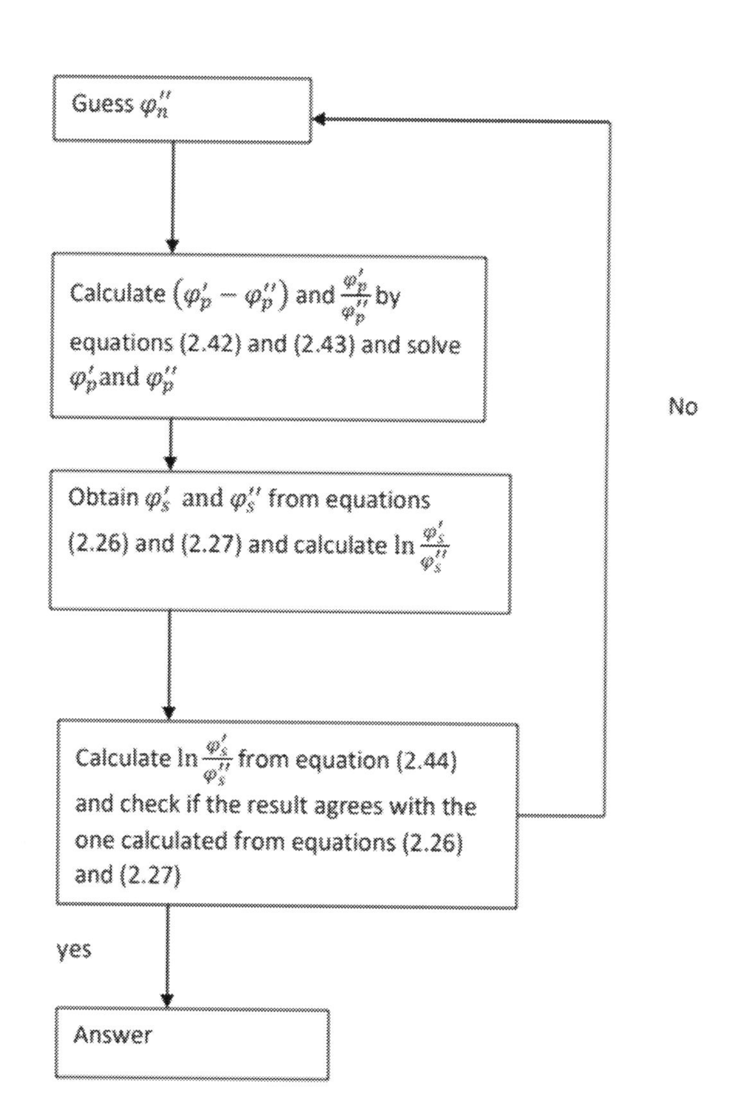

Scheme 2.1 Calculation algorithm used for problem 1.

The results of the calculation are summarized in Table 2.1 and the binodial line is shown in Fig. 2.6.

Problem 2:

Yilmaz and McHugh derived Eqs. (2.45) and (2.46) to know the composition path on the triangular diagram during the phase inversion process (Yilmaz and McHugh, 1986).

$$\varphi_n = \frac{\rho_2^0 - \rho_i^*}{\rho_2^0 - k\rho_1^0} - \frac{\rho_2^0}{\rho_2^0 - k\rho_1^0}\varphi_p \tag{2.45}$$

$$\rho_i^* = (1 + k\omega_{1i})\rho_i \tag{2.46}$$

The definition of the symbols used in the equations is as follows:

φ: volume fraction in the ternary mixture;

Table 2.1 The results of the calculation.

φ_n'	φ_s'	φ_p'	φ_n''	φ_s''	φ_p''
0.38	0.60	0.02	0.38	0.6	0.02
0.4	0.590	0.010	0.350	0.600	0.050
0.5	0.50	0.00	0.290	0.545	0.165
0.7	0.30	0.00	0.223	0.4000	0.377
0.9	0.10	0.00	0.167	0.168	0.665

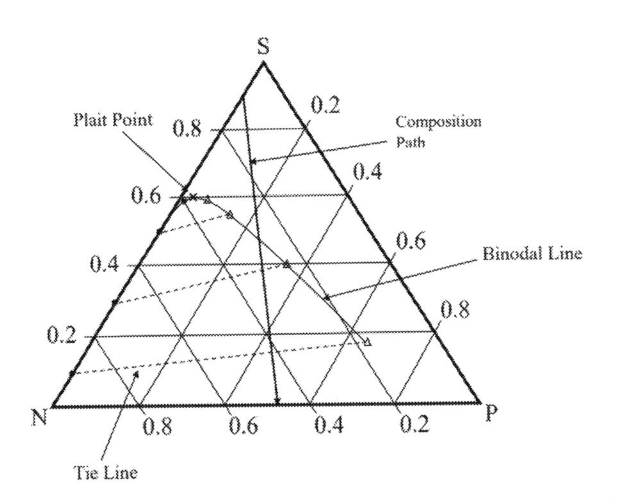

Figure 2.6 Answer to problem 1.

ρ: density defined as the weight of nonsolvent–solvent mixture (free from polymer) in a unit volume of cast film;

$\acute{k}$: ratio of solvent and nonsolvent flux;

$k := \acute{k} - 1$;

ω: mass fraction in nonsolvent–solvent (free from polymer mixture).

Subscripts:

i: initial, when solvent and nonsolvent exchange starts to occur;
n: nonsolvent in ternary mixture;
p: polymer in ternary mixture;
1: nonsolvent in nonsolvent–solvent (free from polymer) mixture;
2: solvent in nonsolvent–solvent (free from polymer) mixture.

Superscript:

0: quantity of pure liquid.

Draw the composition path on Fig. 2.6 when

Initial composition of the casting solution is $\varphi_{si} = 0.9$ and $\varphi_{pi} = 0.1$.

The ratio of solvent and nonsolvent flux $\acute{k} = 1.5$.

The nonsolvent (water) density ρ_1^0 and solvent density ρ_2^0 are 1.0 and 0.79×10^{-3} kg/m^3, respectively.

Answer:

At the start of solvent/nonsolvent exchange, there is only solvent in the cast film. Therefore $\omega_{1i} = 0$ and $\omega_{2i} = 1.0$.

The mass of solvent in unit volume of the cast polymer solution is $\rho_i\omega_{2i}$ at the start of solvent/nonsolvent exchange, which is equal to $\varphi_{si}\rho_2^0$. Therefore

$$\rho_i = \frac{\varphi_{si}\rho_2^0}{\omega_{2i}} = \frac{0.9 \times 0.79}{1,0} = 0.711$$

Then, from Eq. (2.46)

$$\rho_i^* = \{1 + (1.5 - 1) \times 0\} \times 0.711 = 0.711$$

Applying Eq. (2.45)

$$\varphi_n = \frac{0.79 - 0.711}{0.79 - 1.5 \times 1.0} - \frac{0.79}{0.79 - 1.5 \times 1.0}\varphi_p = -0.11127 + 1.1127\varphi_p$$

This is a straight line on the triangular diagram with ($\varphi_{ni} = 0$, $\varphi_{si} = 0.9$ and $\varphi_{pi} = 0.1$) on the P–S axis at one end and ($\varphi_{ne} = 0.474$, $\varphi_{se} = 0$ and $\varphi_{pe} = 0.526$) at the other end.

Nomenclature

Symbols [definition dimension (in SI units)]

m_i	Ratio of molar volume of component i and solvent
n_i	Number of moles of component i
$\boldsymbol{R}$	Gas constant (J/mol K)
$\boldsymbol{T}$	Temperature (K)
$\boldsymbol{x}$	Mole fraction
Greek letters	
ΔG_m	Free energy of mixing (J/mol)
$\Delta\mu$	Change in chemical potential accompanying mixing (J/mol)
φ	Volume fraction
χ	Interaction constant
Superscripts	
$'$	Phase 1
$''$	Phase 2
Subscripts	
$\boldsymbol{n}$	Nonsolvent
$\boldsymbol{p}$	Polymer
$\boldsymbol{s}$	Solvent

References

Matsuura, T., 1994. Synthetic Membranes and Membrane Separation Processes. CRC Press, Florida, Chapter 3).

Tompa, H., 1956. Polymer Solutions. Butterworths, London.

Yilmaz, L., McHugh, A.J., 1986. Development of a diffusion equation formalism for the quench period. J. Membr. Sci. 28, 287–310.

3

Reverse osmosis, forward osmosis, and pressure-retarded osmosis

3.1 Reverse osmosis

3.1.1 Reverse osmosis performance

When the aqueous solutions of two different salt concentrations are separated by a semipermeable membrane, which allows the transport of solvent but does not allow the transport of salt, there is a natural tendency for water to flow from the solution of the lower concentration to the solution of the higher concentration. The driving force for the solvent flow is the difference in osmotic pressure. This phenomenon is called osmosis (Fig. 3.1A).

However, when a hydraulic pressure that is higher than the osmotic pressure is applied on the solution of the higher salt concentration, the direction of the flow is reversed. This phenomenon is called reverse osmosis (RO) (Fig. 3.1C).

The semipermeable membrane is often not perfect and a small amount of salt diffuses from the higher salt concentration to the lower salt concentration.

According to the solution-diffusion model, the RO transport is given by

$$J_A = A(\Delta p - \Delta\pi) \tag{3.1}$$

where J_A is solvent (mostly water) flux, Δp and $\Delta\pi$ are the difference in hydraulic and osmotic pressure, respectively, between both sides of the semipermeable membrane, and the difference, Δ, is defined as (right side–left side in Fig. 3.1). In Eq. (3.1) $\Delta p - \Delta\pi$ is, therefore, considered as the driving force for the water flow from the right to left side. A is a proportionality constant called the water permeation coefficient.

As for solute,

$$J_B = B\Delta c_B \tag{3.2}$$

Membrane Separation Processes. DOI: https://doi.org/10.1016/B978-0-12-819626-7.00006-5

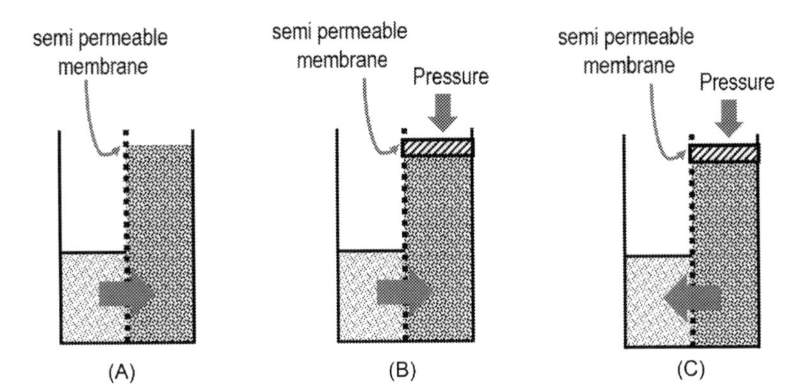

Figure 3.1 (A) Forward osmosis, (B) pressure-retarded osmosis, and (C) reverse osmosis.

where J_B is the solute flux and Δc is the difference in concentration between both sides of the membrane. Again the difference Δ is defined as (from right side to left side). Therefore Δc is always positive and the solute flux is also from right to left. B is a constant called the solute permeation constant.

Furthermore, Lonsdale et al. have shown that

$$J_A = \frac{c_{\mathrm{Am}} D_{\mathrm{Am}} \nu_A}{RT\delta}(\Delta p - \Delta\pi) \tag{3.3}$$

where c_{Am} is the concentration of water in the membrane, D_{Am} is the diffusion coefficient of water in the membrane, ν_A is the molar volume of water, and δ is the membrane thickness (Lonsdale, 1966).

Also,

$$B = \frac{D_{\mathrm{Bm}} K_B}{\delta} \tag{3.4}$$

where D_{Bm} is the diffusion coefficient of solute in the membrane, and K_B is the distribution constant of solute between water and membrane.

In RO the important performance parameters are the solvent flux, which is given by Eq. (3.3), and the solute separation, f', defined as

$$f' = 1 - \frac{c_{\mathrm{B3}}}{c_{\mathrm{B2}}} \tag{3.5}$$

where c_{B2} and c_{B3} are the solute concentration at the high pressure side (i.e., the right side in Fig. 3.1C) and the low pressure side (i.e., the left side in Fig. 3.1C).

The solute separation can be further given by

$$f' = \frac{1}{1 + \frac{D_{Bm}K_B RT c_{A3}}{D_{Am}c_{Am}\nu_A(p_2 - p_3 - \pi_2 + \pi_3)}} \tag{3.6}$$

Problem 3.1:

The following data were given by Lonsdale (1966).
$D_{Am}c_{Am} = 2.7 \times 10^{-8}$ kg/m s and $D_{Bm}K_B = 4.2 \times 10^{-14}$ m^2/s.
Calculate the water flux and the solute separation of sodium chloride based on molality, when the feed sodium chloride molality is 0.1 and the operating pressure, $\Delta p = p_2 - p_3$, is 4.134×10^6 Pa.
The thickness of the membrane is 10^{-7} m.
Use the following numerical values: $RT = 2.479 \times 0^3$ J/mol at 25°C, and $\nu_A = 18.02 \times 10^{-6}$ m^3/mol. The coefficient for the osmotic pressure $= 2.5645 \times 10^8$ Pa per mole fraction.

Answer:

The molality of sodium chloride is 0.1, which means that 0.1 mol of NaCl is dissolved in 1 kg of water. Hence, the mole fraction of NaCl is

$$\frac{0.1}{0.1 + \left(\frac{1000}{18.02}\right)} = 1.799 \times 10^{-3}$$

The osmotic pressure (Pa) is

$$\left(2.5645 \times 10^8\right) \times \left(1.799 \times 10^{-3}\right) = 0.461 \times 10^6$$

Iteration is necessary to calculate the solute separation and flux.
First, solute concentration in the permeate is assumed to be zero.
Therefore $\pi_2 - \pi_3 = 0.461 \times 10^6$ Pa
From Eq. (3.6)

$$f' = \left[1 + \frac{\left(4.2 \times 10^{-14}\right)\left(2.479 \times 10^3\right)\left(10^3\right)}{\left(2.7 \times 10^{-8}\right)\left(18.02 \times 10^{-6}\right)\left(4.134 \times 10^6 - 0.461 \times 10^6\right)}\right]^{-1} = 0.954$$

Then the solute molality in the permeate becomes

$$0.1 \times (1 - 0.945) = 0.0055$$

(It should be noted that the concentration in Eq. (3.6) is kmol/m^3. However, molality is used here because molality is nearly equal to kmol/m^3 in this concentration range.)
The mole fraction of the permeate is

$$\frac{0.0055}{0.0055 + \frac{1000}{18.02}} = 9.910 \times 10^{-5}$$

The osmotic pressure (Pa) of the permeate is

$$(2.5645 \times 10^8) \times (9.910 \times 10^{-5}) = 0.0254 \times 10^6$$

$$\pi_2 - \pi_3 = (0.461 - 0.0254) \times 10^6 = 0.4356 \times 10^6$$

Using the osmotic pressure newly obtained

$$f' = \left[1 + \frac{(4.2 \times 10^{-14})(2.479 \times 10^3)(10^3)}{(2.7 \times 10^{-8})(18.02 \times 10^{-6})(4.134 \times 10^6 - 0.4356 \times 10^6)}\right]^{-1} = 0.945$$

$f' = 0.945$ is therefore accurate enough.
The water flux (kg/m^2 s) is from Eq. (3.3)

$$J_A = \frac{(2.7 \times 10^{-8})(18.02 \times 10^{-6})(4.134 \times 10^6 - 0.4356 \times 10^6)}{(2.479 \times 10^3)(10^{-7})} = 72.56 \times 10^{-4}$$

When there is no solute in the feed, there is no osmotic pressure effect. Therefore

$$J_A = \frac{(2.7 \times 10^{-8})(18.02 \times 10^{-6})(4.134 \times 10^6)}{(2.479 \times 10^3)(10^{-7})} = 81.14 \times 10^{-4}$$

The calculation algorithm is given in Scheme 3.1.

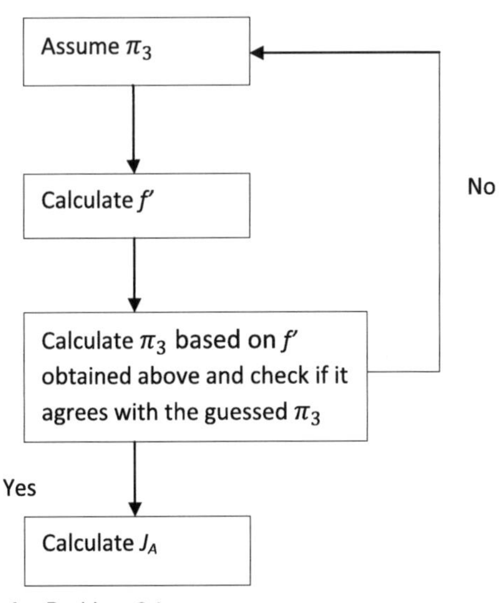

Scheme 3.1 Calculation algorithm for Problem 3.1.

3.2 Concentration polarization

When water permeates through the membrane preferentially from the feed to the permeate, the salt is left behind near the membrane on the feed side unless salt diffuses back to the main body of the feed solution. This phenomenon is called concentration polarization, which causes negative effects on membrane performance such as flux and selectivity reduction. According to the boundary layer theory, concentration polarization is described as follows (Matsuura, 1994).

First, the presence of the boundary layer of thickness, δ_{bl}, is assumed so that the salt diffusion from the membrane to the main body of the feed stream occurs in the boundary layer (see Fig. 3.2; note that the water flow is reversed in Fig. 3.2, that is, water flows from left to right). When the mass balance between the plane at a distance y and the membrane wall at the permeate side is considered,

$$-D_{BA}\frac{dc_B}{dy} + vc_B = vc_{B3} \tag{3.7}$$

where D_{BA} is the diffusion coefficient (m^2/s) of solute B in solvent A (mostly water) in the boundary layer, c_B is the solute concentration, and v is the solution velocity.

The first and second terms of the left-hand side of the equation are the diffusive and convective flow of the solute into a plane at the distance y and the right-hand side is the solute flow from the permeate side of the membrane. They should be equal at the steady state.

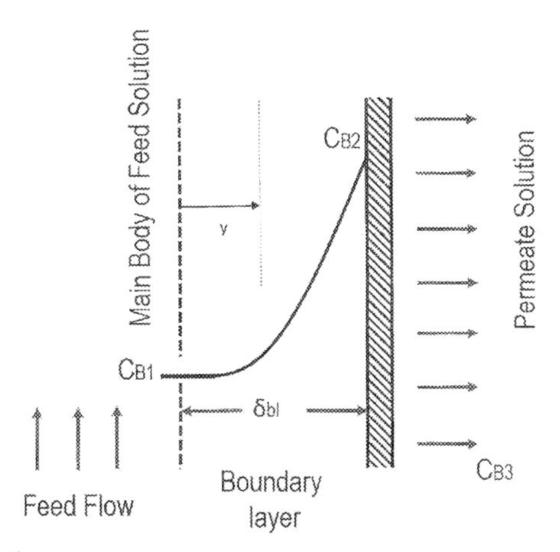

Figure 3.2 Concentration polarization.

Rearranging Eq. (3.7)

$$\frac{dc_B}{dy} = \frac{v}{D_{BA}}(c_B - c_{B3}) \tag{3.8}$$

Then,

$$\frac{dc_B}{c_B - c_{B3}} = \frac{v}{D_{BA}}dy \tag{3.9}$$

Integrating

$$\ln(c_B - c_{B3}) = \frac{v}{D_{BA}}y + C \tag{3.10}$$

where C is the integral constant.

Since $c_B = c_{B1}$ at $y = 0$ (see Fig. 3.2)

$$\ln(c_{B1} - c_{B3}) = C \tag{3.11}$$

Substituting in Eq. (3.11) for Eq. (3.10)

$$\ln\frac{c_B - c_{B3}}{c_{B1} - c_{B3}} = \frac{v}{D_{BA}}y \tag{3.12}$$

Since $c_B = c_{B2}$ at $y = \delta_{bl}$ (see Fig. 3.2)

$$\ln\frac{c_{B2} - c_{B3}}{c_{B1} - c_{B3}} = \frac{v}{D_{BA}}\delta_{bl} \tag{3.13}$$

Defining the mass transfer coefficient as

$$k = \frac{D_{BA}}{\delta_{bl}} \tag{3.14}$$

Eq. (3.13) becomes

$$\ln\frac{c_{B2} - c_{B3}}{c_{B1} - c_{B3}} = \frac{v}{k} \tag{3.15}$$

The boundary concentration, c_{B2}, cannot be obtained experimentally but can be calculated using

Eq. (3.15) by knowing c_{B1}, c_{B3}, v, and k. c_{B1}, c_{B3}, and v are known experimentally and k is often evaluated by dimension analysis.

It should be remembered that the solute separation, f', was defined as

$$f' = 1 - \frac{c_{B3}}{c_{B2}} \tag{3.5}$$

It is impossible to obtain f' experimentally, since c_{B2} cannot be known by experiment. f' can be obtained only by using

Eq. (3.15) through which c_{B2} can be calculated. Another solute separation

$$f = 1 - \frac{c_{B3}}{c_{B1}} \tag{3.16}$$

is used more often. In Eq. (3.16) c_{B1} is known experimentally when the feed solution is prepared. It should be noted however, that f and f' are not equal.

3.3 Prediction of RO performance considering concentration polarization

Prediction of RO performance considering the concentration polarization was attempted by Kimura and Sourirajan (1967). Unlike Lonsdale's derivation that is based on weight-based concentration (kg/m^3) and flux (kg/m^2 s), Kimura–Sourirajna's equations are based on molar concentration (mol/m^3) and molar flux (mol/m^2 s). However, other than those differences, the equations similar to Eqs. (3.1) and (3.2) are used.

From Section 3.2 it is now clear that the solute concentration at the feed solution/membrane interface, called the boundary concentration (c_{B2}), is different from that of the main body of the feed, called the bulk feed concentration, c_{B1}. Hence, from now on, the subscripts 1, 2, and 3 are used for the bulk feed, the boundary, and the permeate throughout the whole book. Since in Eq. (3.1) Δ means the difference between feed solution/membrane interface, 2, and permeate solution/membrane interface, 3, the equation can now be rewritten as

$$J_A = A(p_2 - p_3 - \pi_2 + \pi_3) \tag{3.17}$$

(Note that pressure does not change from the bulk feed to the feed solution/membrane interface, hence $p_1 = p_2$.)

Similarly, the solute flux is written as

$$J_B = B(c_{B2} - c_{B3}) \tag{3.18}$$

Furthermore,

$$c_{B1} = c_1 X_{B1} \tag{3.19}$$

$$c_{B2} = c_2 X_{B2} \tag{3.20}$$

$$c_{B3} = c_3 X_{B3} \tag{3.21}$$

where c is the total molar concentration including solvent and solute and X_B is the mole fraction of the solute.

Substituting Eqs. (3.20) and (3.21) for c_{B2} and c_{B3} in Eq. (3.18),

$$J_B = B(c_2 X_{B2} - c_3 X_{B3}) \tag{3.22}$$

Also using the relation

$$\frac{J_B}{J_A + J_B} = X_{B3} \tag{3.23}$$

$$J_A = B\left(\frac{1 - X_{B3}}{X_{B3}}\right)(c_2 X_{B2} - c_3 X_{B3}) \tag{3.24}$$

Using Eq. (3.15) for concentration polarization, and assuming

$$c_1 = c_2 = c_3 = c \tag{3.25}$$

since the molar concentration of water is much greater than the salt molar concentration in the aqueous solution, and also with the relation

$$v = \frac{J_A + J_B}{c} \tag{3.26}$$

Eq. (3.15) is rearranged to

$$J_A = ck(1 - X_{B3})\ln\frac{X_{B2} - X_{B3}}{X_{B1} - X_{B3}} \tag{3.27}$$

Problem 3.2:

Under the following RO experimental conditions,
Feed: Aqueous NaCl solution;
Feed molality: 0.6;
Operating pressure: 10,335 kPa gage;
Effective membrane area: 13.2×10^{-4} m^2.
The following data were obtained:
Pure water permeation rate: 159.8×10^{-3} kg/h;
Permeation rate for the feed NaCl solution: 122.9×10^{-3} kg/h;
Solute separation based on molality: 81.2%.

Calculate parameters A, B, and k, using the following numerical values,

$$c_1 = c_2 = c_3 = c = 55.3 \text{ kmol/m}^3 \tag{3.28}$$

Molecular weight of NaCl = 58.45 kg/kmol.

Answer:

The flux of pure water is

$$J_A = \frac{(159.8 \times 10^{-3})}{(18.02)(13.2 \times 10^{-4})(3600)} = 1.867 \times 10^{-3}\ \text{kmol/m}^2\,\text{s}$$

In Eq. (3.17), π_2 and π_3 are equal to zero, therefore

$$A = 1.867 \times 10^{-3}/10{,}355 = 1.806 \times 10^{-7}\ \text{kmol/m}^2\,\text{s kPa}$$

As for the flux for the NaCl feed solution,
The permeate molality is

$$(0.6)(1 - 0.812) = 0.1128\,\text{molal}$$

(It should be noted that the solute separation is based on molality instead of kg/m^3 or molarity used for the definition of solute separation by Eq. (3.5) or (3.16). However, the difference between the numerical values of molality and molarity is very small.)

0.1128 mol of NaCl ($0.1128 \times 0^{-3} \times 58.45 = 6.593 \times 10^{-3}$ kg) is in 1 kg of water, then in 122.9×0^{-3} kg of the permeate, the amount of water is $122.9 \times (1/(1 + 6.593 \times 10^{-3})) = 122.1 \times 10^{-3}$ kg.

Therefore water flux is

$$J_A = \frac{(122.1 \times 10^{-3})}{(18.02)(13.2 \times 10^{-4})(3600)} = 1.426 \times 10^{-3}\ \text{kmol/m}^2\text{s}$$

From Table 3.1 the osmotic pressure that corresponds to the permeate molality of 0.1128 molal is 520 kPa.

Table 3.1 Osmotic pressure data pertinent to different electrolyte solutions (at 25°C, kPa).

Molality	NaCl	LiCl	KNO_3	$MgCl_2$	$CuSO_4$
0	0	0	0	0	0
0.1	462	462	448	641	276
0.2	917	931	862	1303	510
0.3	1372	1407	1262	1999	731
0.4	1820	1889	1648	2737	945
0.5	2282	2386	2020	3523	1165
0.6	2744	2889	2379	4357	1379
0.7	3213	3413	2737	5233	1593
0.8	3682	3944	3082	6178	1813
0.9	4158	4482	3427	7191	2055
1.0	4640	5040	3750	8266	2302
1.2	5612	6191	4385	10611	2834

From Eq. (3.17),

$$\pi_2 = p_2 - p_3 + \pi_3 - \frac{J_A}{A} \tag{3.29}$$

Inserting numerical values,

$$\pi_2 = 10,355 + 520 - \frac{1.426 \times 10^{-3}}{1.806 \times 10^{-7}} = 2957 \text{ kPa}$$

From Table 3.1 the molality at the feed solution–membrane interface is 0.6459.

Therefore the mole fractions are

$$X_{B1} = \frac{0.6}{0.6 + \frac{1000}{18.02}} = 0.01070$$

$$X_{B2} = \frac{0.6459}{0.6457 + \frac{1000}{18.02}} = 0.01150$$

$$X_{B3} = \frac{0.1128}{0.1128 + \frac{1000}{18.02}} = 0.002029$$

Rearranging with Eq. (3.24) with the approximation, Eq. (3.28),

$$B = \frac{J_A}{c\left[(1 - X_{B3})/X_{B3}\right](X_{B2} - X_{B3})} \tag{3.30}$$

Inserting numerical values,

$$B = \frac{(1.426 \times 10^{-3})}{(55.3)((1 - 0.002029)/(0.002029))(0.01150 - 0.002029)}$$
$$= 5.536 \times 10^{-6} \text{ m/s}$$

Rearranging Eq. (3.27)

$$k = \frac{J_A}{c(1 - X_{B3})\ln\frac{X_{B2} - X_{B3}}{X_{B1} - X_{B3}}} \tag{3.31}$$

Inserting numerical values,

$$k = \frac{(1.426 \times 10^{-3})}{(55.3)(1 - 0.002029)\ln\left[\frac{(0.01150 - 0.002029)}{(0.01070 - 0.002029)}\right]} = 292.8 \times 10^{-6} \text{ m/s}$$

Problem 3.3:

For a given set of parameters,

$$A = 3.04 \times 10^{-7} \text{ kmol/m}^2\text{ s kPa}$$

$$B = 8.03 \times 10^{-7} \text{ m/s}$$

$$\text{k} = 22 \times 10^{-6} \text{ m/s}$$

calculate the solute separation, f, pure water flux, and permeate flux when the feed is 0.6 molal NaCl solution and the operating pressure is 6895 kPa (gage). Assume that Eq. (3.25) is valid and the osmotic pressure is proportional to NaCl mole fraction.

Answer:

From Eqs. (3.17) and (3.24) under the assumption (Eq. 3.25)

$$A(p_2 - p_3) - A\pi^o(X_{B2} - X_{B3}) = Bc\left[(1 - X_{B3})/X_{B3}\right](X_{B2} - X_{B3}) \tag{3.32}$$

where π^o is the proportional constant between π and X_B.

Rearranging,

$$X_{B2} - X_{B3} = \frac{A(p_2 - p_3)}{A\pi^o + Bc\left[(1 - X_{B3})/X_{B3}\right]} \tag{3.33}$$

From Eqs. (3.17) and (3.27)

$$A(p_2 - p_3) - A\pi^o(X_{B2} - X_{B3}) = ck(1 - X_{B3})\ln\frac{X_{B2} - X_{B3}}{X_{B1} - X_{B3}} \tag{3.34}$$

Inserting the numerical values,

$$A(p_2 - p_3) = (3.04 \times 10^{-7})(6895 - 0) = 20,961 \times 10^{-7}\ \text{kmol/m}^2\,\text{s}$$

which is the pure water permeation flux.

Since the osmotic pressure of 0.6 molal NaCl solution ($X_{B1} = 0.0107$) is 2744 kPa (see Table 3.1)

$$\pi^o = \frac{(2744)}{(0.0107)} = 256,449\ \text{kPa}$$

Therefore

$$A\pi^o = (3.04 \times 10^{-7})(256,449) = 779,600 \times 10^{-7}\ \text{kmol/m}^2\,\text{s}$$

Furthermore,

$$Bc = (8.03 \times 10^{-7})(55.3) = 444.06 \times 10^{-7}\ \text{kmol/m}^2\,\text{s}$$

$$kc = (22 \times 10^{-6})(55.3) = 12,166 \times 10^{-7}\ \text{kmol/m}^2\,\text{s}$$

Inserting the above numerical values in Eq. (3.33)

$$X_{B2} - X_{B3} = \frac{(20,961 \times 10^{-7})}{(779,600 \times 10^{-7}) + (444.06 \times 10^{-7})\left[(1 - X_{B3})/X_{B3}\right]} \tag{3.35}$$

Also, inserting the above numerical values in Eq. (3.34)

$$
\begin{aligned}
&20,961 \times 10^{-7} - 779,600 \times 10^{-7}(X_{B2} - X_{B3}) \\
&\quad = \left(12,166 \times 10^{-7}\right)(1 - X_{B3})\ln\frac{X_{B2} - X_{B3}}{X_{B1} - X_{B3}}
\end{aligned}
\tag{3.36}
$$

Solving Eqs. (3.35) and (3.36) for two unknowns X_{B3} and $X_{B2}{-}X_{B3}$, $X_{B3} = 0.00107$ and $X_{B2}{-}X_{B3} = 0.01755$

Then, $f = \frac{0.0107 - 0.00107}{0.0107} = 0.90$

$$
\begin{aligned}
J_A &= A\left(p_2 - p_3\right) - A\pi^o(X_{B2} - X_{B3}) \\
&= \left(20,961 \times 10^{-7}\right) - \left(779,600 \times 10^{-7}\right)(0.01755) \\
&= 7280 \times 10^{-7}\ \text{kmol/m}^2\ \text{s}
\end{aligned}
$$

3.4 Pore models

3.4.1 Preferential sorption-capillary flow model

According to Sourirajan's book, the following fundamental equation called the Gibbs adsorption isotherm was the basis for the earliest development of RO membrane at the University of California Los Angeles (UCLA) (Sourirajan, 1970).

In Fig. 3.3 an interface is shown between two phases, one the shaded phase, representing air, and the other the unshaded phase, representing the NaCl solution. High up and distant from the interface the solution becomes the bulk solution, whose concentration is c_{Bb}. However, near the interface the concentration c_B is below c_{Bb}. Such an abrupt change of NaCl

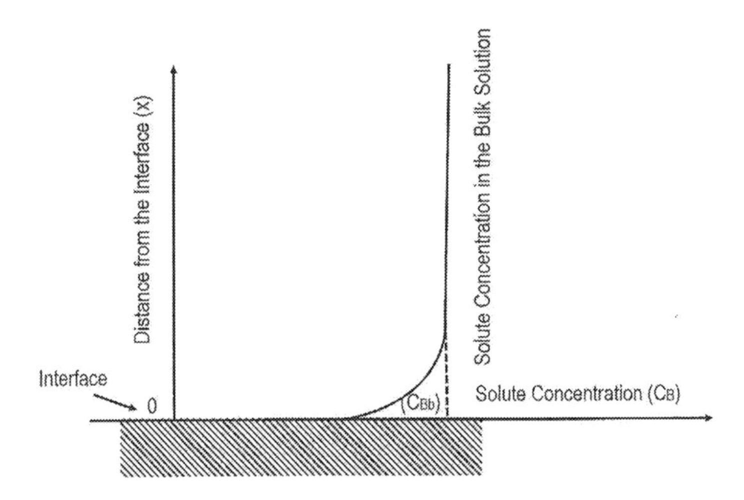

Figure 3.3 Solute concentration profile at the interface showing negative adsorption.

concentration at the interface is predicted by the Gibbs Adsorption Isotherm,

$$\Gamma = -\frac{1}{RT}\frac{\partial \sigma}{\partial \ln a} \tag{3.37}$$

where R is the universal gas constant, T is absolute temperature, σ is surface tension, and a is activity.

Γ is the surface excess given by

$$\Gamma = \int_0^{\infty} (c_B - c_{Bb})dx \tag{3.38}$$

where x is the distance from the interface.

These equations predict the presence of a very thin pure water layer at the surface of NaCl.

Problem 3.4:

Activity coefficient, density, and interfacial tension of aqueous NaCl solutions at 20°C are given for different molalities in Table 3.2. Calculate the interfacial pure water thickness using the data in Table 3.2.

Answer:

Modification of Eq. (3.37) is necessary for the electrolyte solution. For the solution of symmetric electrolytes,

$$a_{\pm} = a^{1/2} \tag{3.39}$$

Combining Eqs. (3.37) and (3.39)

$$\Gamma = -\frac{1}{2RT}\left(\frac{\partial \sigma}{\partial a_{\pm}}\right)_{T,A} = -\frac{1}{2RT}\left(\frac{\partial \sigma}{\partial(\ln(\alpha m_{\pm}))}\right)_{T,A} = -\frac{\alpha m_{\pm}}{2RT}\left(\frac{\partial \sigma}{\partial(\alpha m_{\pm})}\right)_{T,A} \tag{3.40}$$

Since

$$c_{Bb} = \frac{1000m}{1000 + 58.45m}\rho \tag{3.41}$$

where c_{Bb} is the bulk molar concentration of sodium chloride (mol/L).

Then,

$$-\frac{\Gamma}{c_{Bb}} = \frac{\alpha(1000 + 58.54m)}{2RT\rho \times 1000}\left(\frac{\partial \sigma}{\partial(\alpha m)}\right)_{T,A} \tag{3.42}$$

Assuming a stepwise concentration profile at the interface, as illustrated in Fig. 3.4, and considering that $-\Gamma$ is equal to the

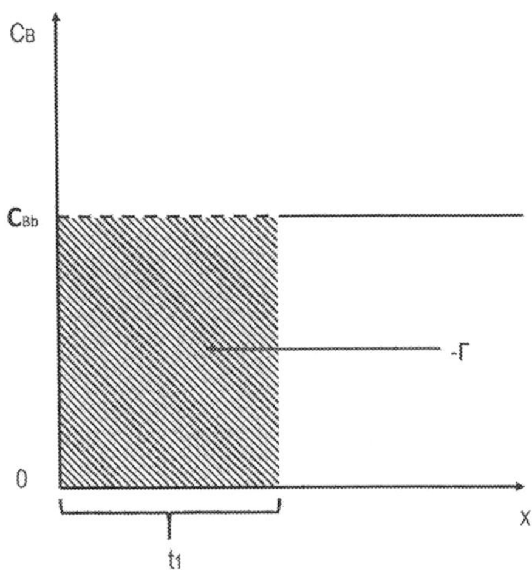

Figure 3.4 Assumption of a stepwise function for the solute concentration profile at the interface.

Table 3.2 Some physicochemical data pertinent to sodium chloride solution.

Molality (m)	Activity coefficient, α	Density, ρ ($\times 10^{-3}$ kg/m^3)	Surface tension, γ ($\times 10^3$ J/m^2)
0.0000	—	—	72.80
0.2010	0.751	1.00675	73.17
0.5030	0.688	1.01876	73.71
1.0204	0.650	1.0385	74.515
2.0988	0.614	1.06984	76.27
3.1920	0.714	1.1125	78.08
4.3628	0.790	1.1507	80.02
4.9730	0.848	1.1679	81.09
5.5410	0.874	1.1947	82.17

shadowed area in the figure, $-\Gamma/c_{Bb}$ is the thickness of the layer where sodium chloride concentration is equal to zero. Hence,

$$t_i = -\frac{\Gamma}{c_{Bb}} \tag{3.43}$$

Table 3.2 was used for the calculation of the pure water thickness, and the results are given in Table 3.3.

According to Sourirajan's preferential sorption-capillary flow model, the pure water layer formed at the salt water–membrane interface is driven by the pressure applied on the feed salty water through subnanometer-sized pores (Fig. 3.5).

Table 3.3 Physicochemical data of sodium chloride solution based on the data given in Table 3.1.

αm (mol/kg)	$\gamma \times 10^3$ (J/m^2)	$\frac{d\gamma}{d(\alpha m)} \times 10^3$	α	$\rho \times 10^{-3}$ (kg/m^3)	m (mol/kg)	$t_i \times 10^{10}$ (m)
0	72.80	2.49a	1.0	1.0	0	5.62
0.5	74.16	2.70	0.669	1.024	0.747	3.78
1.0	75.50	2.52	0.624	1.056	1.603	3.35
1.5	76.68	2.15	0.616	1.081	2.435	2.87
2.0	77.65	1.82	0.640	1.103	3.125	2.57
2.5	78.50	1.67	0.685	1.122	3.650	2.54
3.0	79.32	1.62	0.745	1.139	4.027	2.68
3.5	80.12	1.58	0.795	1.152	4.403	2.82
4.0	80.90	1.49	0.833	1.164	4.802	2.79
4.5	81.61	1.35b	0.861	1.179	5.226	2.64

[a]Calculated by $-3 \times 72.80 + 4 \times 74.16 - 75.50$.
[b]Calculated by $80.12 - 4 \times 80.9 + 3 \times 81.61$.

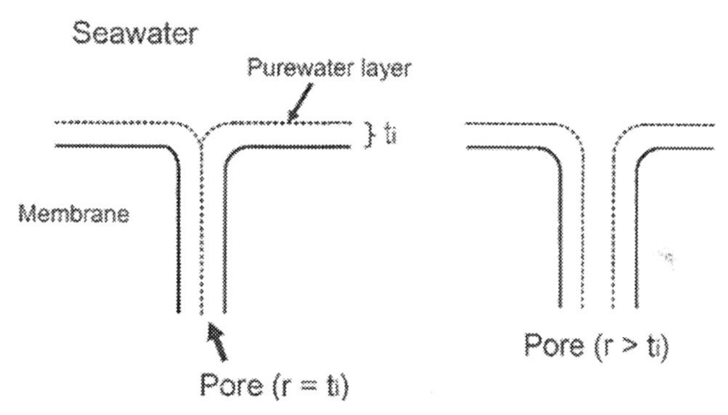

Figure 3.5 Preferential sorption-capillary flow model.

3.4.2 Glückauf model

There are also a number of papers where the RO transport is discussed assuming the presence of pores. One of those is the Glückauf model (Glückauf, 1965).

Suppose the water phase of dielectric constant D (dimensionless) and the polymer phase of dielectric constant D' are in contact with each other and there is a pore of radius r in the polymer phase. When an ion enters the pore, the potential of the ion steadily increases and it reaches a maximum value at the mean distance of the ionic cloud, $1/K$, according to the Debye–Hückel model (Fig. 3.6). When this distance is exceeded, an ion of the opposite charge will enter the pore, reducing the potential of the first ion due to the ion-pair formation. The work required to bring the ionic particle to a distance of $1/K$ from the pore entrance, $\Delta W''$, was approximated by the work required to bring the ion into the cavity of spherical shape shown in Fig. 3.7 and was given by

$$\Delta W'' = \frac{\mathrm{N}Z^2\epsilon^2}{8\pi D(8.854\times 10^{-12})}\frac{(1-\alpha)Q}{r+\alpha bQ} \tag{3.44}$$

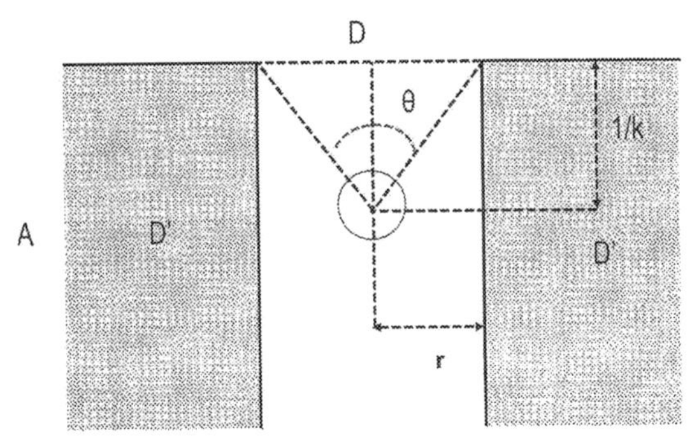

Figure 3.6 Ion is at the distance $1/K$ from the pore entrance.

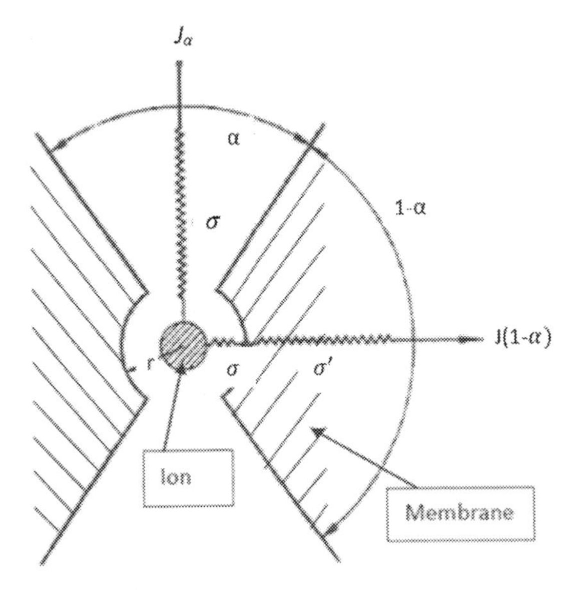

Figure 3.7 Glückauf model.

where Q is D/D', α is the fraction of solid angles over the whole sphere, as shown in Fig. 3.7, which can be given by

$$\alpha = 1 - \left(1 + \kappa^2 r^2\right)^{-1/2} \tag{3.45}$$

and b is the ionic radius.

The probability of finding the ion at this energy level is $\exp(-\Delta W''/RT)$. Thus the concentration in the pore is $c_{B2}\exp(-\Delta W''/RT)$. ($c_{B2}$ is the salt concentration near the feed solution–membrane interface.) Assuming the concentration in the pore is equal to the permeate concentration, c_{B3},

$$c_{B3} = c_{B2}\exp\left(-\frac{NZ^2\epsilon^2}{8\pi D(8.854 \times 10^{-12})RT}\frac{(1-\alpha)Q}{r + \alpha bQ}\right). \tag{3.46}$$

Problem 3.5:

1. Given the following numerical values, calculate the solute separation for pore sizes 0.3, 0.5, 0.7, and 1.0 nm using the Glückauf model for the feed NaCl solution of 1 mol/L.
2. Calculate the solute separation of NaCl for the pore size of 0.5 nm when the feed NaCl concentration is 0.5 mol/L.
3. Calculate the solute separation of $MgSO_4$ for the pore size of 0.5 nm when the $MgSO_4$ concentration is 1.0 mol/L.

Avogadro number $N = 6.023 \times 10^{23}\ \text{mol}^{-1}$
Valence for Na^+ and $Cl^- = 1$, for Mg^{2+} and $SO_4^{2-} = 2$
Electric charge $\varepsilon = 1.602 \times 10^{-19}$ C
Dielectric constant of water $D = 78.54$ at 25°C
An average of dielectric constant of cellulose acetate $D' = 3.7$
Gas constant $R = 8.314$ J/K mol
Absolute temperature $T = 298.2$K
Average of ionic radii of Na^+ and Cl^- $b = 0.142$ nm
Average of ionic radii of Mg^{++} and SO^{4--} $b = 0.1525$ nm
$1/K$ is given as

$$\frac{1}{K} = \sqrt{\frac{Dk_BT}{2\rho_0 N\varepsilon^2}}I^{-1/2} \tag{3.47}$$

where k_B is the Boltzmann constant and ρ_0 is density of water.

I is the ionic strength given as

$$I = \frac{1}{2}\sum_i c_i Z_i^2 \tag{3.48}$$

where c_i and Z_i are ionic concentration and ionic valence, respectively.

$$\frac{1}{K} = 3.05 \times 10^{-10} I^{-1/2} \tag{3.49}$$

can be used instead of Eq. (3.47).

Answer:

(1) For NaCl 1 mol/L solution

$$I = \frac{1}{2}\left(1 \times 1^2 + 1 \times 1^2\right) = 1 \text{ mol/L}$$

From Eq. (3.49)

$$\frac{1}{K} = 3.05 \times 10^{-10} \times 1^{-1/2} = 3.05 \times 10^{-10} \text{ m}$$

When the pore radius is 0.3 nm $\left(= 3 \times 10^{-10} \text{ m}\right)$

$$\alpha = 1 - \left(1 + \left(\frac{3}{3.05}\right)^2\right)^{-1/2} = 0.2871$$

Inserting all numerical values in Eq. (3.46)

$$\frac{c_{B3}}{c_{B2}} = \exp\left(-\frac{\left(6.023 \times 10^{23}\right)\left(1^2\right)\left(1.602 \times 10^{-19}\right)^2}{(8)(3.1416)(78.54)\left(8.854 \times 10^{-12}\right)(8.314)(298.2)}\right.$$

$$\left.\times \left(\frac{(1 - 0.2871)\left(\frac{78.54}{3.7}\right)}{\left(3 \times 10^{-10}\right) + (0.2871)\left(1.42 \times 10^{-10}\right)\left(\frac{78.54}{3.7}\right)}\right)\right) = 0.00972$$

$$f' = 1 - \frac{c_{B3}}{c_{B2}} = 0.99023$$

The remaining answers for problem (1), problem (2), and problem (3) are listed in Table 3.4.

The answers show the following trend.

1. When pore size increases, solute separation decreases.
2. When the solute concentration decreases, solute separation increases.
3. When the ionic valence increases, the solute separation increases.

All the above trends are experimentally observed. However, the increase of solute separation through the decrease of solute concentration seems too large.

Table 3.4 Solute separation calculated for different salts, salt concentrations, and pore sizes.

Solute	Solute concentration (mol/L)	Pore radius $\times 10^{10}$ (m)	f'
NaCl	1	3	0.9902
NaCl	1	5	0.8684
NaCl	1	7	0.6994
NaCl	1	10	0.5058
NaCl	0.5	5	0.9593
$MgSO_4$	1	5	0.9391

Thus the Glückauf model allows prediction of the solute separation when the ionic size, ionic valence, pore size, and dielectric constant of the membrane material are known.

3.5 Forward osmosis and pressure-retarded osmosis

3.5.1 Principles of forward osmosis, reverse osmosis, and pressure-retarded osmosis

The principles of forward osmosis (FO), RO, and pressure-retarded osmosis (PRO) are shown in Fig. 3.1, where the solution of the lower solute concentration and the higher solute concentration are placed on the left- and right-hand sides, respectively, of a semipermeable membrane. Thus the osmotic pressure of the solution is higher on the right than the left.

The equation that describes the relationship between water flux and hydraulic and osmotic pressures is given for all the above processes by

$$J_A = A(\Delta p - \Delta\pi) \tag{3.50}$$

where J_A is water flux, Δp and $\Delta\pi$ are the differences in hydraulic and osmotic pressure, respectively, between both sides of the semipermeable membrane. It should be remembered that the difference, Δ, is defined as (right side−left side), hence $\Delta\pi$ is always positive. In Eq. (3.1), $\Delta p - \Delta\pi$ is considered as the driving force for the water flux from the right to left side. A is a proportionality constant called the water permeation coefficient.

In FO, there is no hydraulic pressure difference on both sides, therefore $\Delta p = 0$. Thus, according to Eq. (3.50), the driving force $\Delta p - \Delta\pi$ is negative and the direction of water flow is from left to right. Since the solution of higher salt concentration placed on the right side draws water from the left, it is often called the draw solution (DS). In RO, pressure is applied on the right side so that $\Delta p - \Delta\pi$ becomes positive. Therefore the flow direction is reversed as from right to left (Fig. 3.8C) in contrast to FO. Because the flow direction is reversed from the natural osmotic flow, it is called RO. In PRO, pressure is applied on the right side but $\Delta p - \Delta\pi$ is maintained as negative, hence the flow direction is from left to right (Fig. 3.8B). Then, the hydraulic energy (flow rate $\times$ pressure) is converted into mechanical or electrical energy by a turbine.

As for the solute flux, J_B, it is given by

$$J_B = B\Delta c \tag{3.51}$$

where Δc is the difference in concentration between both sides of the membrane. Again the difference Δ is defined as (right − left). Therefore Δc is always positive. B is a constant called the solute permeation constant.

3.5.2 Applications of forward osmosis

Some examples of FO applications are as follows.

Emergency bag: The solution of ingestible solute, for example, solution of high sugar concentration, is loaded in a bag made of semipermeable membrane. When the bag is placed in a surface water (pond, lake, etc.) a sufficient amount of potable water permeates into the bag in a sufficient amount of time, leaving undesirable constituents of the surface water. Thus the diluted sugar solution in the bag can be drunk.

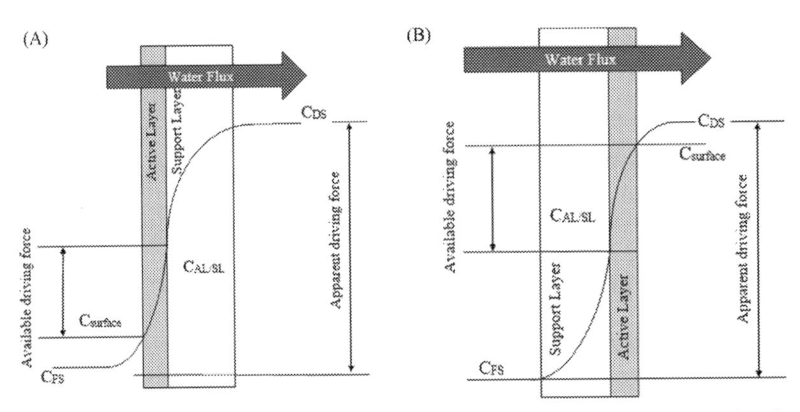

Figure 3.8 Schematic presentation of the salt concentration profile in FO: (A) active layer facing feed water (support layer facing draw solution) and (B) support layer facing feed water (active layer facing draw solution).

Desalination: Desalination of seawater or brackish water is possible using a solution of high osmotic pressure as DS. The advantage of this process is that high feed pressure is not required and membrane fouling is less than RO. However, another process is required to separate the permeated water from the DS, which can be achieved by other membrane separation processes, thermal processes, physical processes, and a combination of these with additional energy requirement.

Evaporative cooling water make-up: Water evaporated during the cooling process is supplied by FO from water sources such as seawater, brackish water, sewage effluent, or industrial wastewater.

FO/RO hybrid process: Water is drawn from landfill leachate or municipal wastewater into saline water as DS. Then, the diluted saline water is subjected to RO to produce potable water as the permeate. The concentrate is reused in FO.

3.5.3 Concentration polarization in forward osmosis

Similar to RO, concentration polarization takes place in FO. There are two kinds of concentration polarization in FO, external and internal. The external concentration polarization is the same as the concentration polarization in RO. It occurs outside the membrane. On the other hand, the internal concentration polarization occurs inside the membrane when a dense selective membrane is supported by a porous substrate membrane (Fig. 3.8).

Fig. 3.8A shows the profile of the salt concentration across the membrane when the dense active layer (AL) is facing the feed solution while the porous support layer (SL) is facing DS, which is called AL-facing-FW (AL–facing-feed water). The feed concentration is lower than the concentration in DS, hence water flows from left to right, as explained earlier. However, because of the external concentration polarization, the salt concentration at the feed–membrane interface is higher than the bulk feed concentration. Also, due to the internal concentration polarization occurring in the SL, the salt concentration at the active (rejection) layer/SL boundary is lower than the concentration in bulk DS. Hence the concentration difference across the AL is much lower than the concentration difference between the bulk feed solution and the bulk DS. This means that the driving force for the FO flux is greatly reduced since the osmotic pressure difference is nearly proportional to the concentration difference.

Fig. 3.8B shows the profile of the salt concentration across the membrane when the dense active (rejection) layer is facing DS while the porous SL is facing the feed solution, which is called AL-facing-DS. Similar to Fig. 3.8A, both internal and external concentration polarization occur to reduce the FO flux.

Comparing Figs. 3.8A and B, the internal concentration is more severe in Fig. 3.8B as the porous SL is facing the DS with a much higher salt concentration.

3.5.4 Forward osmosis transport

The following FO transport theory is based on the work by Tang et al. (2010).

From the RO experiment the water A and solute permeation constant B, respectively, are obtained by the following equations.

$$A = \frac{J_{\text{water}}}{\Delta P} \tag{3.52}$$

$$\frac{1-f'}{f'} = \frac{B}{A(\Delta P - \Delta\pi)} \tag{3.53}$$

where f' is the salt rejection.

First, AL-facing-DS (Fig. 3.8B) is discussed.

Applying Eq. (3.50) for the dense AL

$$J_A = A\left(\pi_{DS} - \pi_{AL/SL}\right) \tag{3.54}$$

$$J_B = B\left(c_{DS} - c_{AL/SL}\right) \tag{3.55}$$

where J_A ($m^3/m^2/s$) is the flux of A (water) from feed water to DS, π_{DS}(kPa) and $\pi_{AL/SL}$ (kPa) are the osmotic pressure of DS and the osmotic pressure of the solution at the AL/SL interface, respectively. [It should be noted that contrary to Eq. (3.50) the driving force is defined as $\Delta\pi$ instead of $-\Delta\pi$, and, accordingly, the flux of water from FS side to DS side (from left to right) is defined positive in Eq. (3.54). The flux of solute from DS to FS is defined positive in Eq. (3.55), which is the same as in Eq. (3.51).]

c (mol/m^3) is the solute concentration and the subscript DS and AL/SL indicate the DS and the solution at the AL/SL interface, respectively.

From the mass balance of the solute in the SL

$$-D_{eff}\frac{dc}{dx} - J_A c = J_B \tag{3.56}$$

where D_{eff} (m^2/s) is the diffusion coefficient of solute in the porous SL, and x is the distance from the AL/SL interface toward FS.

The first term $-D_{\text{eff}}(dc/dx)$ represents the diffusive flow of salt toward the bulk feed and the second term $-J_A c$ is the convective flow of salt toward the AL, and J_B on the right-hand side is the solute transport through the AL.

Rearranging Eq. (3.56)

$$-D_{eff}\frac{dc}{dx}=J_A c+J_B \tag{3.57}$$

$$\frac{-dc}{c+\frac{J_B}{J_A}}=\frac{J_A dx}{D_{eff}} \tag{3.58}$$

Integrating

$$-\ln\left(c+\frac{J_B}{J_A}\right)=\frac{J_A x}{D_{eff}}+\text{Const} \tag{3.59}$$

Since

$$c=c_{AL/SL}\quad \text{at}\quad x=0 \tag{3.60}$$

$$\text{Const}=-\ln\left(c_{AL/SL}+\frac{J_B}{J_A}\right) \tag{3.61}$$

From Eqs. (3.59) and (3.61)

$$-\ln\left(\frac{c+\frac{J_B}{J_A}}{c_{AL/SL}+\frac{J_B}{J_A}}\right)=\frac{J_A x}{D_{eff}} \tag{3.62}$$

Since

$$c=c_{FS}\quad \text{at}\quad x=l_{eff}=\tau l \tag{3.63}$$

$$-\ln\left(\frac{c_{FS}+\frac{J_B}{J_A}}{c_{AL/SL}+\frac{J_B}{J_A}}\right)=\frac{J_A\tau l}{D_{eff}} \tag{3.64}$$

Rearranging

$$\ln\left(\frac{c_{AL/SL}+\frac{J_B}{J_A}}{c_{FS}+\frac{J_B}{J_A}}\right)=\frac{J_A\tau l}{D_{eff}} \tag{3.65}$$

Inserting Eqs. (3.54) and (3.55)

$$\ln\left(\frac{c_{AL/SL}+B(c_{DS}-c_{AL/SL})/A(\pi_{DS}-\pi_{AL/SL})}{c_{FS}+B(c_{DS}-c_{AL/SL})/A(\pi_{DS}-\pi_{AL/SL})}\right)=\frac{J_A}{k_m} \tag{3.66}$$

where

$$k_m=\frac{D_{eff}}{l_{eff}}=\frac{D}{\tau l/\varepsilon}=\frac{D}{S} \tag{3.67}$$

and ε is the porosity.

The structural parameter, S, analogous to the thickness of the external boundary layer, is defined as

$$S = \frac{\tau l}{\varepsilon} \tag{3.68}$$

Assuming that osmotic pressure is proportional to the solute concentration, Eq. (3.66) is simplified as

$$\ln\left(\frac{\pi_{AL/SL} + B/A}{\pi_{FS} + B/A}\right) = \frac{J_A}{k_m} \tag{3.69}$$

From Eqs. (3.54) and (3.69)

$$J_A = k_m \ln\left(\frac{A\pi_{DS} - J_A + B}{A\pi_{FS} + B}\right) \tag{3.70}$$

for AL-facing-DS modes.

Similarly, for AL-facing-FW

$$\ln\left(\frac{c_{DS} + B(c_{AL/SL} - c_{FS})/A(\pi_{AL/SL} - \pi_{FS})}{c_{AL/SL} + B(c_{AL/SL} - c_{FS})/A(\pi_{AL/SL} - \pi_{FS})}\right) = \frac{J_A}{k_m} \tag{3.71}$$

and

$$J_A = k_m \ln\left(\frac{A\pi_{DS} + B}{A\pi_{FS} + J_A + B}\right) \tag{3.72}$$

Problem 3.6:

From the RO experiments with a thin-film nanocomposite membrane (TFN0.25), the following data were obtained (Emadzadeh et al., 2014).

What are A and B?

The water flux was 5.475 $L/m^2\,h$ and solute separation was 93.7%, at the feed sodium chloride concentration of 1000 ppm and operating pressure of 2.5 bar (250 kPa) gage.

Assume that the osmotic pressure is proportional to the sodium chloride mole fraction with a proportionality constant 256,449 kPa/mole fraction.

Answer:

From Eq. (3.52)

$$A = \frac{\left(\frac{5.475 \times 10^{-3}}{3600}\right)}{250} = 6.083 \times 10^{-9}\ \mathrm{m^3/m^2\,s\,kPa}$$

The mole fraction of NaCl in the 1000 ppm NaCl (c. 1 kg/m^3) solution is

$$\frac{\left(\frac{1000}{58.45}\right)}{\left\{\left(\frac{1000}{58.45}\right)+\left(\frac{1{,}000{,}000-1000}{18.02}\right)\right\}}=0.0003085$$

Feed osmotic pressure is $0.0003085 \times 256,449 = 79.11$ kPa

Permeate osmotic pressure is $0.0003085 \times (1-0.937) \times 256,449 = 4.98$ kPa

Hence, $\Delta\pi = 79.11 - 4.98 = 74.13$ kPa

From Eq. (3.53)

$$B=\frac{1-f'}{f'}A(\Delta P-\Delta\pi)=\frac{(1-0.937)}{0.937}(6.083\times 10^{-9})(250-74.13)$$

$$=71.93\times 10^{(-9)}\ \mathrm{m^3/m^2\,s}$$

Problem 3.7:

Using the same membrane, fluxes of 36 and 22 $L/m^2\,h$ were obtained for AL-facing-DS and AL-facing-FS, respectively, by the FO experiments with pure water feed and DS of 2 M NaCl concentration. Calculate k_m, assuming that the osmotic pressure is proportional to mole fraction and the density of 2 M solution is 1.08 kg/L. What then is the mole fraction of NaCl at the AL/SL interface?

Answer:

The mole fraction of 2 M sodium chloride in DS is

$$\frac{(2)}{\left(\frac{(1000\times 1.08-2\times 58.45)}{(18.02)}+2\right)}=0.03607$$

The osmotic pressure is $0.03607 \times 256{,}449 = 9250$ kPa

From Eq. (3.70)

$$\frac{J_A}{k_m}=\ln\left(\frac{(6.083\times 10^{-9})(9250)-\left(36\times\frac{10^{-3}}{3600}\right)+(71.93\times 10^{-9})}{(71.93\times 10^{-9})}\right)=6.468$$

Therefore $k_m = \dfrac{\left(36\times\frac{10^{-3}}{3600}\right)}{6.468} = 1.546\times 10^{-6}$ m/s

From Eq. (3.54)

$$\pi_{AL/SL}=\pi_{DS}-\frac{J_A}{A}=(9250)-\frac{\left(36\times\frac{10^{-3}}{3600}\right)}{(6.083\times 10^{-9})}=7606\ \mathrm{kPa}$$

The mole fraction at the support is then,

$$\frac{(7606)}{(256,449)}=0.0297$$

which is equal to 1.669 M.

For AL-facing-FS, from Eq. (3.72),

$$\frac{J_A}{k_m} = \ln\left(\frac{(6.083\times10^{-9})(9250)+(71.93\times10^{-9})}{\left(22\times\frac{10^{-3}}{3600}\right)+(71.93\times10^{-9})}\right) = 2.210$$

$$k_m = \frac{\left(22\times\frac{10^{-3}}{3600}\right)}{2.210} = 2.765\times10^{-6}\ \text{m/s}$$

k_ms of AL-facing-DS and AL-facing-FS are supposed to be the same, but they do not agree. Why?

In AL-facing-FS, FS (in this case pure water) is facing the AL instead of DS, therefore

$$\pi_{AL/SL} = \pi_{FS} + \frac{J_A}{A} = \frac{\left(22\times\frac{10^{-3}}{3600}\right)}{(6.083\times10^{-9})} = 1005\ \text{kPa}$$

The mole fraction at the AL/porous support interface is then,

$$\frac{(1005)}{(256,449)} = 0.0039$$

Which corresponds to 0.232 M.

Problem 3.8:

Using the same membrane in AL-facing-DS mode, what is the flux when the NaCl concentration in DS is 1, 3, and 4 mol/L and the feed is pure water? Use densities of 1.05, 1.12, and 1.15 kg/L, respectively, for 1, 3, and 4 mol/L.

Answer:

For AL-facing-DS:

When the DS concentration is 1 mol/L, the mole fraction is

$$\frac{(1)}{\left(\frac{(1000\times1.05-1\times58.45)}{(18.02)}+1\right)} = 0.01785$$

Osmotic pressure of DS, π_{draw} is

$$0.01785\times256,449 = 4578\ \text{kPa}$$

Solving Eq. (3.70) by iteration

$$J_A = 8.64\times10^{-6}\ \text{m}^3/\text{m}^2\ \text{s}$$

When DS is 3 mol/L, the mole fraction is

$$\frac{(3)}{\left(\frac{(1000\times1.12-3\times58.45)}{(18.02)}+3\right)} = 0.05413$$

Osmotic pressure of DS, π_{draw} is

$$0.05413 \times 256,449 = 13,881 \text{ kPa}$$

$$J_A = 10.72 \times 10^{-6} \text{ m}^3/\text{m}^2 \text{ s}$$

When DS is 4 M, the mole fraction is

$$\frac{(4)}{\left(\frac{(1000 \times 1.15 - 4 \times 58.45)}{(18.02)} + 4\right)} = 0.07293$$

Osmotic pressure of DS, π_{draw} is

$$0.07293 \times 256,449 = 18,702 \text{ kPa}$$

$$J_A = 11.23 \times 10^{-6} \text{ m}^3/\text{m}^2 \text{ s}$$

The flux versus DS concentration is given in Fig. 3.9.

Problem 3.9:

Calculate the power that can be generated by PRO as a function of the pressure applied on DS under the following conditions.

Membrane: TFN 0.25
Membrane area: 1 m^2
FS: pure water
NaCl concentration of DS: 3 mol/L
AL-facing-DS

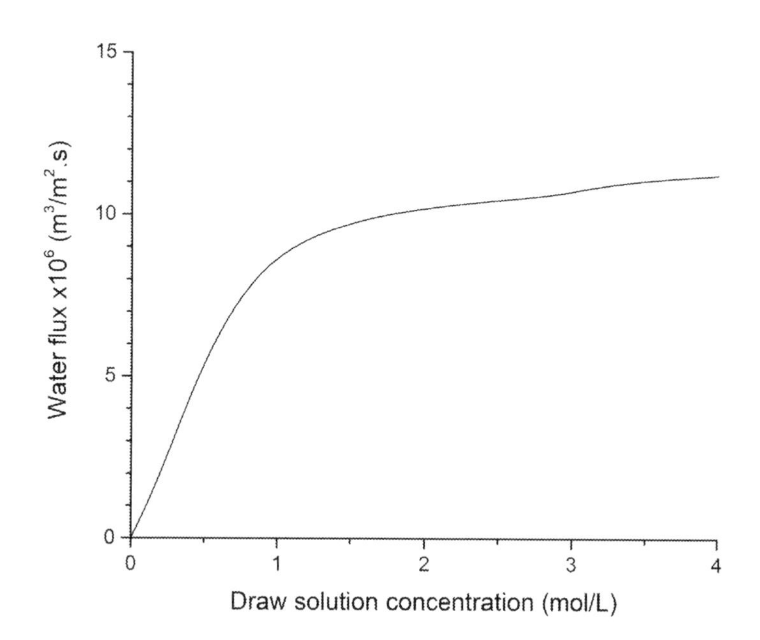

Figure 3.9 Flux versus concentration (AL-facing-DS).

Answer:

Instead of Eq. (3.54), the following equation should be used.

$$J_A = A(\pi_{DS} - \pi_{AL/SL} - \Delta p) \tag{3.73}$$

where $\Delta p = p_{DS} - p_{FS}$

Then, from Eq. (3.66)

$$\ln\left(\frac{C_{AL/SL} + B(C_{DS} - C_{AL/SL})/A(\pi_{DS} - \pi_{AL/SL} - \Delta p)}{C_{FS} + B(C_{DS} - C_{AL/SL})/A(\pi_{DS} - \pi_{AL/SL} - \Delta p)}\right) = \frac{J_A}{k_m} \tag{3.74}$$

Assuming osmotic pressure is proportional to NaCl concentration,

$$\ln\left(\frac{\pi_{AL/SL} + B(\pi_{DS} - \pi_{AL/SL})/A(\pi_{DS} - \pi_{AL/SL} - \Delta p)}{\pi_{FS} + B(\pi_{DS} - \pi_{AL/SL})/A(\pi_{DS} - \pi_{AL/SL} - \Delta p)}\right) = \frac{J_A}{k_m} \tag{3.75}$$

Using Eq. (3.73)

$$\ln\left(\frac{A\pi_{DS} - A\Delta p - J_A + B\left(\frac{J_A + A\Delta p}{J_A}\right)}{A\pi_{FS} + B\left(\frac{J_A + A\Delta p}{J_A}\right)}\right) = \frac{J_A}{k_m} \tag{3.76}$$

J_A can be solved from Eq. (3.76) by iteration.

Then, the power E can be given by

$$E = \Delta p J_A = \Delta p k_m \ln\left(\frac{A\pi_{DS-A\Delta p-J_A} + B\left(\frac{J_A + A\Delta p}{J_A}\right)}{A\pi_{FS} + B\left(\frac{J_A + A\Delta p}{J_A}\right)}\right) \tag{3.77}$$

The osmotic pressure of 3 mol/L sodium chloride solution is 13,881 kPa. J_A can be solved from Eq. (3.76) by iteration and E is given by Eq. (3.77). The results are listed in Table 3.5 for different Δp values.

Table 3.5 J_A and E for different Δp values.

P (kPa)	$J_A \times 10^6$ (m^3/m^2 s)	E (W/m^2)
4631	7.70	35.65
9303	5.10	47.45
11,000	3.74	41.14
13,000	0.69	8.97

Nomenclature

Symbol	**[definition, dimension (SI units)]**
$\boldsymbol{a}$	Activity (molality)
$\boldsymbol{b}$	Ionic radius (m)
A	Water permeation coefficient (kg/m^2 s Pa in Eq. 3.1, kmol/m^2 s Pa in Eq. 3.17)
B	Solute permeation constant (m/s)
$\boldsymbol{c}$	Total molar concentration including solute and solvent (water) (kmol/m^3)
c_B	Concentration of solute (kg/m^3 in Eq. 3.2, kmol/m^3, in Eq. 3.18)
c_{Bb}	Bulk molar concentration (kmol/m^3)
c_i	Ionic concentration (kmol/m^3)
$\boldsymbol{c_{\mathrm{Am}}}$	Concentration of water in the membrane (kg/m^3)
$\boldsymbol{D}$	Water phase dielectric constant
$\boldsymbol{D'}$	Polymer phase dielectric constant
D_{Am}	Diffusion coefficient of water in the membrane (m^2/s)
D_{Bm}	Diffusion coefficient of solute in the membrane (m^2/s)
D_{BA}	Diffusion coefficient of solute A in solvent B (m^2/s)
$\boldsymbol{f}$	Solute separation based on bulk feed concentration
f'	Solute separation based on boundary feed concentration
I	Ionic strength (kmol/m^3)
J_A	Solvent (mostly water) flux (kg/m^2 s in Eq. 3.1, kmol/m^2 s in Eq. 3.17)
J_B	Solute flux (kg/m^2 s in Eq. 3.2, kmol/m^2 s in Eq. 3.18)
k	Mass transfer coefficient (m/s)
K_B	Distribution constant of solute between water and membrane
$\boldsymbol{m}$	Molarity (kmol/kg of solvent)
$\boldsymbol{N}$	Avogadro number (1/mol)
$\boldsymbol{p}$	Pressure (Pa)
$\boldsymbol{r}$	Pore radius (m)
$\boldsymbol{R}$	Gas constant (J/mol K)
$\boldsymbol{T}$	Temperature (K)
$\boldsymbol{v}$	Solution velocity (m/s)
X_B	Mole fraction of solute
$\boldsymbol{Z}$	Ionic valence
Z_i	Valence for ith ion
Greek letters	
α	Activity coefficient in Eq. (3.40), fraction of solid angle in Eq. (3.45)
Γ	Surface excess (kmol/m^2)
δ	Membrane thickness (m)
$\Delta W''$	Work required to bring the ionic particle to distance of κ^{-1} from pore entrance (J/mol)
$\boldsymbol{\delta_b}$	Boundary layer thickness (m)
ε	Electric charge (C)
κ^{-1}	Mean distance of ionic cloud (m)
ν_A	Molar volume of water (m^3/mol)
π	Osmotic pressure (Pa)
ρ	Density (kg/m^3)
σ	Surface tension (J/m^2)
Subscripts	
1	Feed, bulk solution
2	Feed, boundary layer
3	Permeate

A	Water permeation coefficient (m^3/m^2 s Pa)
B	Solute permeation constant (m/s)
c	Solute concentration (kg/m^3)
$c_{AL/SL}$	Solute concentration at the active layer/support layer interface (kg/m^3)
c_{DS}	Solute concentration of draw solution (kg/m^3)
D_{eff}	Diffusion coefficient of solute in the porous support (m^2/s)
J_A	Volumetric flux of water (m^3/m^2 s)
J_B	Solute flux (kg/m^2 s)
f'	Salt rejection
k_m	Mass transfer coefficient (m/s)
l	Membrane thickness (m)
l_{eff}	Effective membrane thickness (m)
p	Pressure (Pa)
S	Structural parameter (m)
Greek letters	
ε	Membrane porosity
π	Osmotic pressure (Pa)
$\pi_{AL/SL}$	Osmotic pressure at the active layer/support layer interface (Pa)
π_{DS}	Osmotic pressure of draw solution (Pa)
τ	Tortuosity factor

References

Emadzadeh, D., Lau, W.J., Ismail, A.F., Matsuura, T., Rahbari-Sisakht, M., 2014. Synthesis and characterization of novel thin film nanocomposite forward osmosis membrane with hydrophilic nanocomposite support to reduce internal concentration polarization. J. Membr. Sci. 449, 74–85.

Glückauf, E., 1965. On the mechanism of osmotic desalting with porous membranes. In: Proceedings, First International Symposium on Water Desalination, vol. 1. US. Department of the Interior, Office of Saline Water, Washington, DC, pp. 143–156.

Kimura, S., Sourirajan, S., 1967. Analysis of data in reverse osmosis with porous cellulose acetate membranes used. AIChE J. 13, 497.

Lonsdale, H.K., 1966. Properties of cellulose acetate membranes. In: Merten, U. (Ed.), Desalination by Reverse Osmosis. MIT Press, Cambridge, Chapter 4).

Matsuura, T., 1994. Synthetic Membranes and Membrane Separation Processes. CRC Press, Boca Raton, FL, Chapter 5).

Sourirajan, S., 1970. Reverse Osmosis. Academic Press, Cambridge, MA.

Tang, C.Y., She, Q., Lay, W.C.L., Wang, R., Fane, A.G., 2010. Coupled effects of internal concentration polarization and fouling on flux behavior of forward osmosis membranes during humic acid filtration. J. Membr. Sci. 354, 123–133.

4

Nanofiltration

4.1 Solution in general

Nanofiltration (NF) is another pressure-driven membrane separation process which is often used for water softening (i.e., separation of divalent and monovalent cations). The pore size of NF membranes is slightly larger than that of reverse osmosis (RO) membranes, ranging from 1 to 10 nm. Since the membrane is often charged, the effect of the membrane charge should be considered in the transport theory. Dresner applied the extended Nernst–Planck equation to calculate the ionic rejections by RO (he called it hyperfiltration) for multicomponent systems (Dresner, 1972). Later Tsuru et al. also used the extended Nernst–Planck equation to calculate the ionic rejection for single- and multicomponent systems (Tsuru et al., 1991). Bowen and Mukhtar further applied the extended Nernst–Planck equation to predict NF performances. In particular, they introduced the hindrance factor for the convective flow to consider the effect of pore size on the membrane performance (Bowen and Mukhtar, 1996; Bowen and Mohammad, 1998).

According to Bowen and Mukhtar, the separation of ions by charged porous membranes can be predicted by knowing the membrane thickness, Δx, membrane charge density, X, and membrane pore size, r_p. Their pore model is schematically illustrated in Fig. 4.1.

The model is written as:

$$J_i = -D_{i,p}\frac{dc_{i,p}}{dx} - \frac{z_i c_{i,p}}{RT}F\frac{d\psi}{dx} + K_{i,c}c_{i,p}J_A \tag{4.1}$$

where J_i is the flux of i-th ion, $D_{i,p}$ is the diffusivity of the ion in the pore, $c_{i,p}$ is the concentration of ions in the pore, R is the gas constant, T is absolute temperature, F is Faraday constant, ψ is electric potential, and $K_{i,c}$ is the hindrance factor for the convective flow. The first, second, and the third terms of the right-hand side of Eq. (4.1) represent the electrical transport due to diffusion, electric field gradient, and convection, respectively.

Membrane Separation Processes. DOI: https://doi.org/10.1016/B978-0-12-819626-7.00008-9

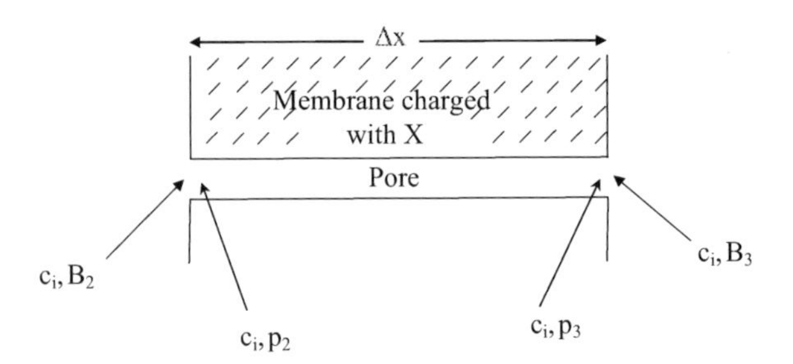

Figure 4.1 Pore model for nanofiltration.

The concentration gradient can be obtained by rearranging Eq. (4.1) as

$$\frac{dc_{i,p}}{dx} = \frac{J_A}{D_{i,p}}(K_{i,c}c_{i.p} - c_{i,3}) - \frac{z_i c_{i,p}}{RT} F \frac{d\psi}{dx} \tag{4.2}$$

and the ion flux is written as

$$J_i = J_A c_{i,3} \tag{4.3}$$

where $c_{i,3}$ is the ion concentration in the permeate. (Later, it is called $c_{cat,B3}$ and $c_{ani,B3}$, for cation and anion, respectively.)

With an assumption that the effective charge density, X, is constant from the pore inlet to the pore outlet, that is,

$$\sum_{i=1}^{n} z_i c_{i,p} = -X \quad (0 \leq x \leq \Delta x) \tag{4.4}$$

where Δx is the pore length.

Multiplying Eq. (4.3) by z_i and adding all ions and applying, $\frac{d\sum_{i=1}^{n} z_i c_{i,p}}{dx} = 0$, the potential gradient becomes

$$\frac{d\psi}{dx} = \frac{\sum_{i=1}^{n} \frac{z_i J_A}{D_{i,p}}(K_{i,c}c_{i.p} - c_{i,3})}{\frac{F}{RT}\sum_{i=1}^{n} z_i^2 c_{i,p}} \tag{4.5}$$

There is also a requirement of electroneutrality in the solution outside the pore,

$$\sum_{i=1}^{n} z_i c_{i,B} = 0 \tag{4.6}$$

where the subscript B means external solution (solution outside the pore), either at the pore entrance or pore outlet.

And there is no electric current

$$I = \sum_{i=1}^{n} F z_i J_i = 0 \tag{4.7}$$

Furthermore, Donnan equilibrium is applied both at the pore entrance and pore exit:

$$\frac{c_{i,p}}{c_{i,B}} = \exp\left(-\frac{z_i F}{RT}\Delta\psi_D \right) \quad \text{at } x = 0 \text{ and } \Delta x \tag{4.8}$$

4.2 Solution for mono-monovalent electrolytes

The differential Eqs. (4.2) and (4.5) can then be solved numerically, for a mono-monovalent electrolyte, as follows.

In the feed solution from Eq. (4.6),

$$c_{cat,B2} = c_{ani,B2} = c_{B2} \tag{4.9}$$

where the subscript *2* means "at the pore entrance."

In the pore from Eq. (4.4),

$$c_{cat,p2} - c_{ani,p2} + X = 0 \tag{4.10}$$

From Eq. (4.8),

$$\frac{c_{cat,p2}}{c_{cat,B2}} = \exp\left(-\frac{F}{RT}\Delta\psi_D \right) \tag{4.11}$$

and

$$\frac{c_{ani,p2}}{c_{ani,B2}} = \exp\left(+\frac{F}{RT}\Delta\psi_D \right) \quad \text{at } x = 0 \tag{4.12}$$

Thus, $c_{cat,p2}$, $c_{ani,p2}$, and $\Delta\psi_D$ can be solved from Eqs. (4.10), (4.11), and (4.12) for a given X.

4.3 Solution method

The initial guess of c_{B3} is made. Eq. (4.6) is now written as

$$c_{cat,B3} = c_{ani,B3} = c_{B3} \tag{4.13}$$

and the differential equations

$$\frac{dc_{cat,p}}{dx} = \frac{J_A}{D_{cat,p}}(K_{cat,c}c_{cat,p} - c_{cat,B3}) - \frac{c_{cat,p}}{RT}F\frac{d\psi}{dx} \tag{4.14}$$

$$\frac{d\psi}{dx} = \frac{\frac{J_A}{D_{cat,p}}(K_{cat,c}\,c_{cat,p} - c_{cat,B3}) - \frac{J_A}{D_{ani,p}}(K_{ani,c}\,c_{ani,p} - c_{ani,B3})}{\frac{F}{RT}(c_{cat,p} + c_{ani,p})} \tag{4.15}$$

Starting from the ionic concentration at the pore entrance [Eq. (4.9)], Eqs. (4.14) and (4.15) are carried out until $c_{cat,p3}$ is reached at $x = \Delta x$. Then, $c_{cat,B3} = c_{B3}$ is calculated by

$$\frac{c_{cat,B3}}{c_{cat,p3}} = \exp\left(\frac{F}{RT}\Delta\psi_D\right) \tag{4.16}$$

It is then used for the second guess of $c_{cat,B3}$. The iteration is continued until the last two guesses are in agreement.

The rejection is then given by

$$f_i' = 1 - \frac{c_{B3}}{c_{B2}} \tag{4.17}$$

Tsuru et al. have given a numerical solution for the differential equations when the solute is mono-monovalent electrolyte and when $K_{i,c'}$s are equal to unity.

The results are as follows.

$$\frac{J_A \Delta x}{\epsilon} = -D_s\left\{\frac{1}{2}\ln\frac{Z(c_{B3})^2 - 2c_{B3}Z(c_{B3}) - A}{Z(c_{B2})^2 - 2c_{B3}Z(c_{B2}) - A} + \frac{c_{B3}}{2B}\ln\left(\frac{Z(c_{B3}) - c_{B3} - B}{Z(c_{B3}) - c_{B3} + B}\right)\left(\frac{Z(c_{B2}) - c_{B3} + B}{Z(c_{B2}) - c_{B3} - B}\right)\right\} \tag{4.18}$$

where

$$A = (-X)^2 + 2(2\alpha - 1)(-X)c_{B3} \tag{4.19}$$

$$B = \left(((-X) - c_{B3})^2 + 4\alpha(-X)c_{B3}\right)^{1/2} \tag{4.20}$$

$$Z(c_{Bi}) = \left(4c_{Bi}^2 + (-X)^2\right)^{1/2} \quad i = 2 \text{ or } 3 \tag{4.21}$$

Eq. (4.18) is rewritten as

$$\frac{J_A \Delta x}{\epsilon} = -D_s\left(\frac{1}{2}\ln P + R\ln Q\right) \tag{4.22}$$

where

$$P = \frac{Z(c_{B3})^2 - 2c_{B3}Z(c_{B3}) - A}{Z(c_{B2})^2 - 2c_{B3}Z(c_{B2}) - A} \tag{4.23}$$

$$Q = \left(\frac{Z(c_{B3}) - c_{B3} - B}{Z(c_{B3}) - c_{B3} + B}\right) \times \left(\frac{Z(c_{B2}) - c_{B3} + B}{Z(c_{B2}) - c_{B3} - B}\right) \tag{4.24}$$

$$R = \frac{c_{B3}}{2B} \tag{4.25}$$

By using $c_{B3} = \gamma c_{B2}$ and $-X = \xi c_{B2}$, Eqs. (4.23) to (4.25) are further written as

$$P = \frac{4\gamma^2 - 2\gamma\sqrt{4\gamma^2 + \xi^2} - 2\alpha(2\alpha - 1)\xi\gamma}{4 - 2\gamma\sqrt{4 + \xi^2} - 2\alpha(2\alpha - 1)\xi\gamma} \tag{4.26}$$

$$Q = \frac{\sqrt{4\gamma^2 + \xi^2} - \gamma - \sqrt{(\xi - \gamma)^2 + 4\alpha\xi\gamma}}{\sqrt{4\gamma^2 + \xi^2} - \gamma + \sqrt{(\xi - \gamma)^2 + 4\alpha\xi\gamma}} \times \frac{\sqrt{4 + \xi^2} - \gamma + \sqrt{(\xi - \gamma)^2 + 4\alpha\xi\gamma}}{\sqrt{4 + \xi^2} - \gamma - \sqrt{(\xi - \gamma)^2 + 4\alpha\xi\gamma}} \tag{4.27}$$

$$R = \frac{\gamma}{2\sqrt{(\xi - \gamma)^2 + 4\alpha\xi\gamma}} \tag{4.28}$$

Using the equations developed by Tsuru et al. the solute separation can be calculated by the following algorithm (Scheme 4.1).

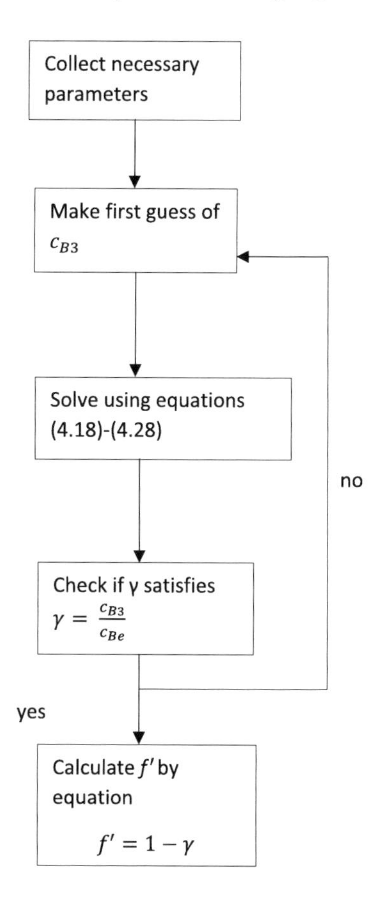

Scheme 4.1 Calculation algorithm for the nanofiltration of mono-monovalent electrolyte.

Problem 4.1:

Calculate the separation of sodium chloride by a NF membrane with a pore radius of 1 nm, porosity 6.9%, thickness 2 μm, and flux of 360 L/m^2 h, for the ratios of membrane charge to the NaCl concentration at the solution–membrane interface ($-X/c_{B2}$) are 0.1, 0.2, . . ., 1. Use the diffusivity of sodium chloride in water = 1.61×10^{-9} m^2/s.

Answer:

(Tsuru et al.'s equation for mono-monovalent salt is used assuming that the diffusivity of the Na^+ and Cl^- ions are governed by those in the pore, and the partition coefficient, $K_{i,c}$, is equal to unity.)

$$J_A = 360 \text{L/m}^2\text{h} = 360(1\ \text{m}^3/1000\text{L})/\text{m}^2(3600\ \text{s}/1\ \text{h}) = 10^{-4}\text{m/s}$$

$$\frac{J_A \Delta x}{\epsilon} = \frac{10^{-4} \times 2 \times 10^{-6}}{0.069} = 2.90 \times 10^{-9}\text{m}^2/\text{s}$$

The diffusion of ions is hindered considerably in the membrane pore and the ratio of the diffusivity in the pore to the diffusivity without hindrance is given by

$$K_{id} = \frac{D_{i,p}}{D_{i,\infty}} \tag{4.29}$$

where $D_{i,p}$ is the diffusivity of ions in the pore and $D_{i,\infty}$ is the diffusivity of ions without hindrance.

Furthermore, K_{id} is given as the function of the ratio of ionic radius to the pore radius as

$$K_{id} = -1.705\lambda + 0.946 \tag{4.30}$$

where

$$\lambda = \frac{r_i}{r_p} \tag{4.31}$$

The results of the calculations are listed in Table 4.1.

$$\alpha = \frac{D_{cat,p}}{D_{cat,p} + D_{ani,p}} = \frac{0.844}{0.844 + 1.503} = 0.360 \tag{4.32}$$

Table 4.1 Properties of sodium and chloride ions.

Ions	$D_{i,\infty} \times 10^9$, m^2/s	$r_i \times 10^9$, m	λ	K_{id}	$D_{i,p} \times 10^9$, m^2/s
Na^+	1.333	0.1840	0.1840	0.633	0.844
Cl^-	2.031	0.1207	0.1207	0.740	1.503

Table 4.2 f' for different ξ values.

ξ	γ	f'
0.1	0.85	0.15
0.2	0.746	0.254
0.3	0.69	0.31
0.4	0.672	0.328
0.5	0.679	0.321
0.6	0.703	0.297
1.0	0.885	0.115

The diffusivity of NaCl in the pore is obtained as the geometric average of the cation and anion as

$$D_p = \frac{2D_{cat,p}D_{ani,p}}{D_{cat,p} + D_{ani,p}} = 1.081 \times 10^{-9}\text{m}^2/\text{s}$$

Then, γ is obtained from Eqs. (4.18)–(4.28) by iteration for different values of ξ, and solute separation f' is equal to $1-\gamma$.

γ and f' are listed for different ξs in Table 4.2.

Nomenclature

Symbols [definition dimension (SI unit)]

c	Concentration (mol/m^3)
D	Diffusivity (m^2/s)
$D_{i,\infty}$	Diffusivity of ion without hindrance (m^2/s)
D_s	Diffusivity of salt in the pore (m^2/s)
f'	Solute separation
F	Faraday constant (C/mol)
I	Electric current (A)
J	Flux (mol/m^2 s)
$K_{i,c}$	Hindrance factor for convective flow
K_{id}	Hindrance factor for diffusion
n	Total number of ions
r_i	Ionic radius (m)
r_p	Pore radius (m)
R	Gas constant (J/mol K)
T	Absolute temperature (K)
x	Distance from the pore entrance (m)
X	Membrane charge (mol/m^3)
z	Valence

Greek letters

α	Quantity defined by Eq. (4.32)
γ	Ratio of permeate concentration to feed concentration

Δx	Pore length (m)
$\Delta\psi_D$	Donnan potential difference (V)
ε	Porosity
λ	Ratio of ionic radius to pore radius
ξ	Ratio of membrane charge density to feed concentration
ψ	Electric potential (V)
Subscripts	
ani	Anion
A	Water
B	Solution outside of pore
cat	Cation
i	*i*-th ion
p	Inside the pore
2	Feed
3	Permeate

References

Bowen, W.R., Mohammad, A.W., 1998. Characterization and prediction of nanofiltration membrane performance-A general assessment. Trans. IChemE 76 (PartA), 885–893.

Bowen, W.R., Mukhtar, H., 1996. Characterization and prediction of separation performance of nanofiltration membranes. J. Membr. Sci 112, 263–274.

Dresner, L., 1972. Some remarks on the integration of the extended Nernst-Planck equations in the hyperfiltration of multicomponent solutions. Desalination 10, 27–46.

Tsuru, T., Nakao, S., Kimura, S., 1991. Calculation of ioninc rejection by extended Nernst Planck equation with charged revere osmosis membranes for single and mixed electrolyte solutions. J. Chem. Eng. Japan 24, 511–517.

5

Ultrafiltration and microfiltration

5.1 Ultrafiltration: gel model

In ultrafiltration (UF) of macromolecular or colloidal solutes, the flux increases initially in a linear fashion as the feed pressure increases, but the flux levels off at high feed pressures as shown in Fig. 5.1 (Jonsson, 1984), Blatt et al. attempted to explain this phenomenon by the gel mode (Blatt et al., 1970).

Similar to the concentration polarization in RO

$$\ln \frac{c_{B2} - c_{B3}}{c_{B1} - c_{B3}} = \frac{v}{k} \tag{5.1}$$

When the solute rejection is almost equal to 100%, $c_{B3} \approx 0$, then

$$\ln \frac{c_{B2}}{c_{B1}} = \frac{v}{k} \tag{5.2}$$

Suppose the feed pressure is gradually increased while maintaining the feed concentration c_{B1} and the mass transfer coefficient k constant, then the permeation velocity v increases due to the driving force increase, and, according to Eq. (5.2), c_{B2} also increases. When c_{B2} reaches a critical value, called the gel concentration, c_{gel}, a gel layer is formed at the membrane surface and c_{B2} can no longer surpass this value. Then, Eq. (5.2) becomes

$$\ln \frac{c_{gel}}{c_{B1}} = \frac{v}{k} \tag{5.3}$$

According to Eq. (5.3) v remains constant, even when pressure is further increased. The flux versus pressure plot then becomes as illustrated in Fig. 5.1. Note that the flux increases linearly when the pressure is low, but gradually deviates from the linear line and eventually levels off at high pressures. From the figure, it is also seen that the flux after the gel formation decreases as c_{B1} increases.

The gel concentration c_{gel} can be evaluated in the following way.

Membrane Separation Processes. DOI: https://doi.org/10.1016/B978-0-12-819626-7.00004-1

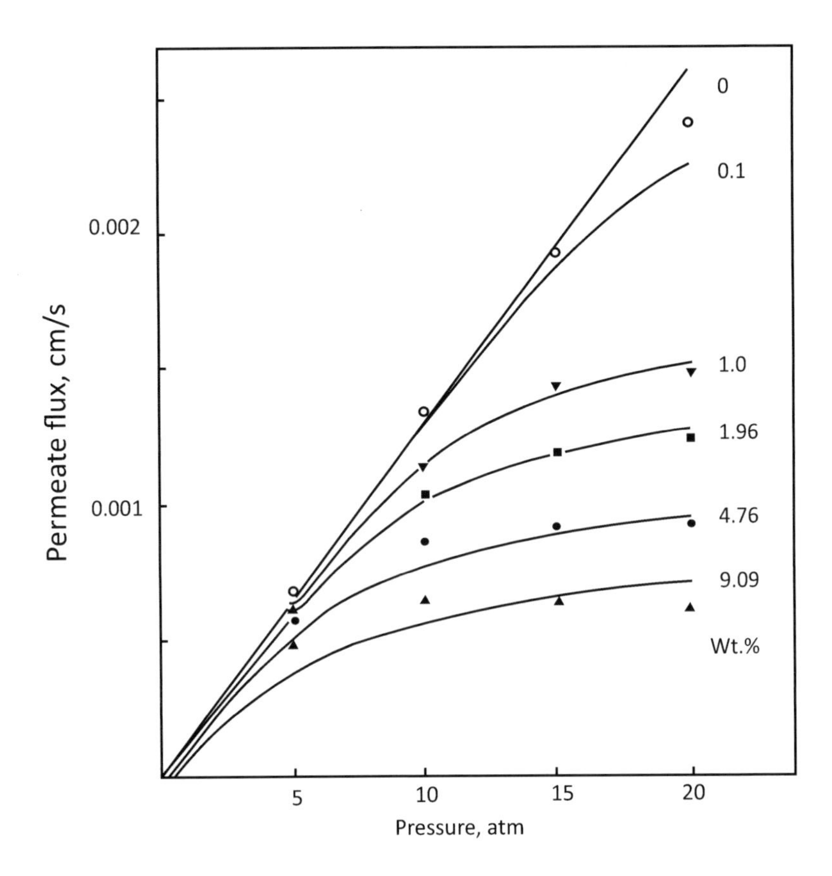

Figure 5.1 Typical flux and pressure relationship of ultrafiltration.

Rearrangement of Eq. (5.3) yields,

$$v = k \ln c_{gel} - k \ln c_{B1} \tag{5.4}$$

Hence, when a linear plot v versus $\ln c_{B1}$ is made, k is obtained from the slope and $v = 0$ when $c_{B1} = c_{gel}$.

Problem 5.1:

Obtain the gel concentration using the data given in Fig. 5.1.

Answer:

From Fig. 5.1, a set of data given in Table 5.1 is obtained.

By applying the least square analysis for the plot y (flux $\times$ 10^2) and x ($\ln c_{B1}$), we obtain

$$y = -0.0004x + 0.001466$$

Then, $x = 3.665$ for $y = 0$ and the gel concentration is $e^{3.665} = 39.06$ wt.%.

The data in Table 5.1 are plotted in Fig. 5.2, together with ($x = 3.665, y = 0$).

Table 5.1 Experimental data for flux versus whey protein concentration, and ln c_{gel} obtained from the linear plot.

Feed protein concentration, c_{B1}, (wt.%)	x: ln c_{B1}	y: Flux × 10^2 (m^3/m^2 s)
1.0	0	0.00145
1.96	0.672	0.0012
4.76	1.560	0.0009
9.09	2.207	0.00015

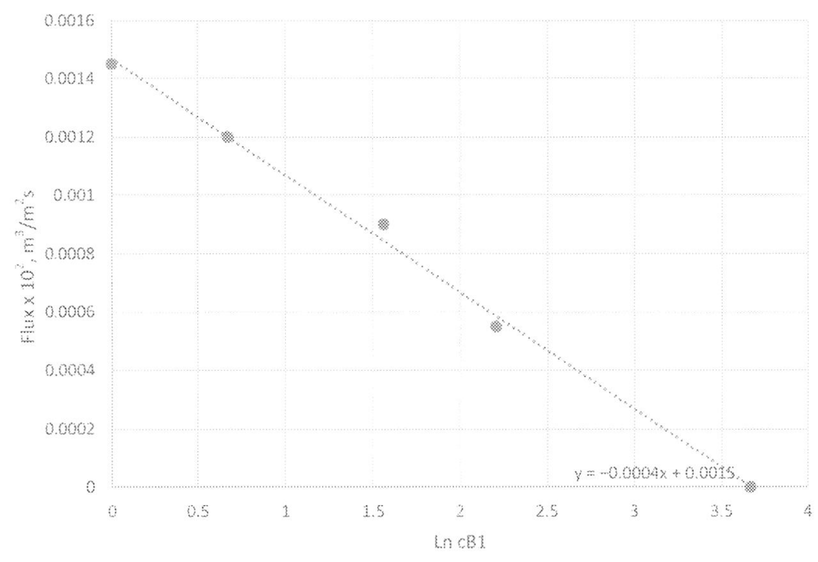

Figure 5.2 Linear plot of flux versus ln c_{B1}.

5.2 Microfiltration: Brownian diffusion, shear-induced diffusion, inertia lift

Microfiltration (MF) is a pressure-driven process by which suspended colloids and particles with sizes of 0.1–20 μm are separated from the filtrate under pressures below 0.35 MPa. Very high fluxes ranging from 10^{-4} to 10^{-2} m^3/m^2 s can be achieved by MF (Belfort et al., 1994).

5.2.1 Brownian diffusion

During the MF operation, a thin cake layer of colloids and a boundary layer above the cake layer are built at the membrane surface. Both the thickness of the cake layer and that of the

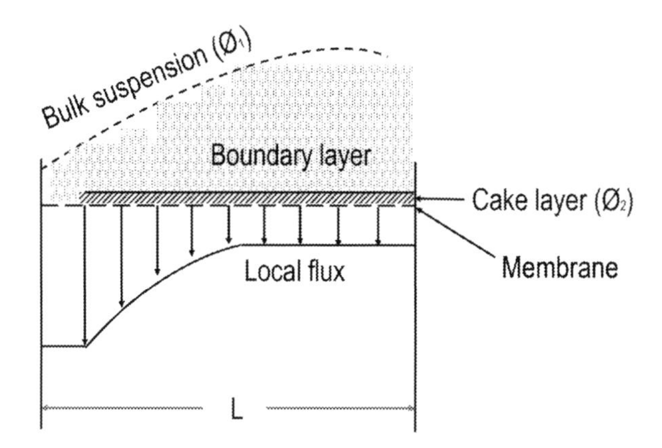

Figure 5.3 Cake layer and boundary layer formation from the edge of a flat sheet membrane.

boundary layer increase with the distance from the edge of the membrane, as shown in Fig. 5.3.

When the steady state is reached in the boundary layer, the convective particle flow toward the membrane surface is balanced by the diffusive flow away from the membrane surface, and, similar to the gel model given in UF, the permeation velocity becomes,

$$v = K \ln \frac{\phi_{B2}}{\phi_{B1}} \tag{5.5}$$

where K is the mass transfer coefficient, and ϕ_{B1} and ϕ_{B2} are the particle volume fraction in the bulk suspension and in the cake layer formed at the membrane surface.

The length-averaged mass transfer coefficient is given by solving the Leveque equation for the boundary layer, which yields

$$K = 0.81\left(\frac{\dot{\gamma}_0 D^2}{L}\right)^{1/3} \tag{5.6}$$

where $\dot{\gamma}_0$ is the shear rate, D is the diffusivity of the particle, and L is the channel length.

The Brownian diffusion coefficient for the isolated particles in the fluid of viscosity η_0 is given by

$$D_{B0} = \frac{kT}{6\pi\eta_0 a} \tag{5.7}$$

by the Stokes–Einstein relationship, where a is the particle size.

Then, combining Eqs. (5.5)–(5.7), we obtain (Belfort et al., 1994)

$$v = 0.114\left(\dot{\gamma}_0 k^2 T^2/{\eta_0}^2 a^2 L\right)^{1/3} \ln \frac{\phi_{B2}}{\phi_{B1}} \tag{5.8}$$

According to Eq. (5.8), the permeation velocity v decreases as a increases and becomes very small when the particle size is more than 1 μm. Experimentally observed values were, however, one or more orders of magnitude higher than the value predicted by the Brownian diffusion coefficient. Hence, this was called the "flux paradox for colloidal suspension."

5.2.2 Shear-induced diffusion

To resolve the flux paradox, Zydney and Colton proposed a concentration polarization model in which the Brownian diffusion coefficient was replaced by the shear-induced diffusivity (Zydney and Colton, 1986). Shear-induced diffusion of particles occurs when individual particles undergo random displacement from the stream lines in a shear flow as they interact with and tumble over other particles. Approximate shear-induced diffusivity, D_s, for $0.2 < \phi_{B1} < 0.45$ is

$$D_s = 0.03\dot{\gamma}_0 a^2 \tag{5.9}$$

Replacing the Brownian diffusion coefficient of Eq. (5.7) with the shear-induced diffusivity

$$v = 0.078\dot{\gamma}_0 \left(a^4/L\right)^{1/3} \ln\frac{\phi_{B2}}{\phi_{B1}} \tag{5.10}$$

It should be noted that, according to Eq. (5.10), the permeation velocity is a to the power of 4/3 and increases as a increases.

5.2.3 Inertial lift

Belfort and coworkers further attempted to solve the paradox by inertial lift working on the colloidal particles (Green and Belfort, 1980). Inertial lift arises from the nonlinear interactions of a particle with the surrounding flow field under conditions where the Reynolds numbers based on the particle size are no longer negligible and the noninertia terms in the Navier–Stokes equation play the role. The inertial lift velocity of spherical particles under the laminar flow conditions in dilute suspensions, where particle–particle interactions are negligible, is given by

$$v_{L0} = \frac{b\rho_0 a^3 \dot{\gamma}_0^{\,2}}{16\eta_0} \tag{5.11}$$

where $b = 0.577$ for fast laminar flow and ρ_0 is the density of the liquid.

Table 5.2 Flux obtained by the equations for Brownian diffusion, shear-induced diffusion, and inertia lift.

a, μm	Brownian diffusion	J_A, m/s Shear induced diffusion	Inertia lift
1	2.395×10^{-7}	1.191×10^{-5}	2.25×10^{-7}
10	0.517×10^{-7}	2.564×10^{-4}	2.25×10^{-4}
50	0.176×10^{-7}	2.191×10^{-3}	2.81×10^{-2}

Then, the permeation velocity becomes

$$v = v_{L0} = 0.036\rho_0 a^3 \dot{\gamma}_0{}^2/\eta_0 \tag{5.12}$$

It should be noted that v is proportional to the third power of a.

Problem 5.2:

Calculate the permeation velocity for the particles with radii of 1, 10, and 50 μm, using the Brownian diffusion, shear-induced diffusion, and inertia lift, respectively. Use the following values:

$$\dot{\gamma}_0 = 2500\ \mathrm{s}^{-1},\ k = 1.38 \times 10^{-23}\mathrm{J\ K}^{-1}, T = 293\ K,$$

$$\eta_0 = 0.001\ \mathrm{kg/m}\ s,\ \rho_0 = 1000\ \mathrm{kg/m}^3,\ \phi_{B1} = 0.01,$$

$$\phi_{B2} = 0.6 \text{ and } L = 0.3\ \mathrm{m}.$$

Answer:

For $a = 10^{-6}$ m,

Brownian diffusion: $v = 0.114 \times \left(\frac{2500 \times (1.38 \times 10^{-23})^2 \times 293^2}{0.001^2 \times (10^{-6})^2 \times 0.3}\right)^{1/3} \times \ln\frac{0.6}{0.01} = 2.395 \times 10^{-7}\mathrm{m/s}$;

Shear-induced diffusion: $v = 0.078 \times (2500) \times \left(\frac{(10^{-6})^4}{0.3}\right)^{1/3} \times \ln\frac{0.6}{0.01} = 1.191 \times 10^{-5}\mathrm{m/s}$;

Inertia lift: $v = 0.036 \times \frac{1000 \times (10^{-6})^3 \times (2500)^2}{0.001} = 2.25 \times 10^{-7}\mathrm{m/s}$.

Similarly, the flux is calculated for 10 and 50 μm. The results are summarized in Table 5.2.

Nomenclature

Symbol [definition dimension (SI unit)]

c_B	Concentration of solute (kg/m^3)
c_{gel}	Gel concentration (kg/m^3)
k	Mass transfer coefficient (m/s)
v	Solution velocity (m/s)

Subscripts

1	Feed, bulk solution
2	Feed, boundary layer
3	Permeate

Symbol [definition dimension (SI unit)]

a	Particle size (m)
D	Diffusivity of particle (m^2/s)
D_{B0}	Brownian diffusion coefficient (m^2/s)
D_s	Shear induced diffusivity (m^2/s)
k	Boltzmann constant (J/K)
K	Mass transfer coefficient (m/s)
L	Channel length (m)
T	Temperature (K)
v	Permeate velocity (m/s)
v_{L0}	Inertial lift velocity of spherical particles under the laminar flow conditions in dilute suspensions (m/s)

Greek letters

$\dot{\gamma}_0$	Shear rate (1/s)
η_0	Fluid viscosity (Pa s)
ρ_0	Fluid density (kg/m^3)
ϕ_B	Particle volume fraction

References

Belfort, G., Davis, R.H., Zydney, A.L., 1994. The behavior of suspensions and macromolecular solutions in crossflow microfiltration. J. Membr. Sci. 96, 1–58.

Blatt, W.F., Dravid, A., Michaels, A.S., Nelson, L., 1970. Solute polarization and cake formation in membrane ultrafiltration: causes, consequences and control techniques. In: Flinn, J.E. (Ed.), Membrane Science and Technology. Plenum Press., New York.

Green, G., Belfort, G., 1980. Fouling of ultrafiltration membranes: lateral migration and the particle trajectory model. Desalination 35, 129–147.

Jonsson, G, 1984. Boundary layer phenomena during ultrafiltration of dextran and whey protein solutions. Desalination 61, 61–77.

Zydney, A.L., Colton, C.K., 1986. A concentration polarization model for the filtration flux in cross-flow microfiltration of particulate suspensions. Chem. Eng. Commun. 47, 1–21.

6

Membrane gas separation

6.1 Solution-diffusion model

6.1.1 Steady-state transport

The solution-diffusion model is most widely used for the gas transport of nonporous membrane. According to the model, gas is dissolved into the membrane on the feed side, diffuses across the membrane and is disengaged from the membrane on the permeate side (see Fig. 6.1).

The flux is given according to Fick's first law as

$$J = -D\frac{dc}{dx} \tag{6.1}$$

where J is gas flux, c is the concentration of gas in the membrane, and x is the distance from the feed side toward the permeate side of the membrane.

Assuming diffusivity, D, is constant throughout the membrane, integration of Eq. (6.1) yields,

$$J = D\frac{(c_2 - c_3)}{\delta} \tag{6.2}$$

where subscripts 2 and 3 indicate the feed and the permeate side of the membrane, respectively.

The concentration in the membrane is given by Henry's law as

$$c = Sp \tag{6.3}$$

where S is solubility coefficient and p is pressure.

Then, from Eqs. (6.2) and (6.3)

$$J = DS\frac{(p_2 - p_3)}{\delta} \tag{6.4}$$

or

$$J = P\frac{(p_2 - p_3)}{\delta} \tag{6.5}$$

Membrane Separation Processes. DOI: https://doi.org/10.1016/B978-0-12-819626-7.00007-7

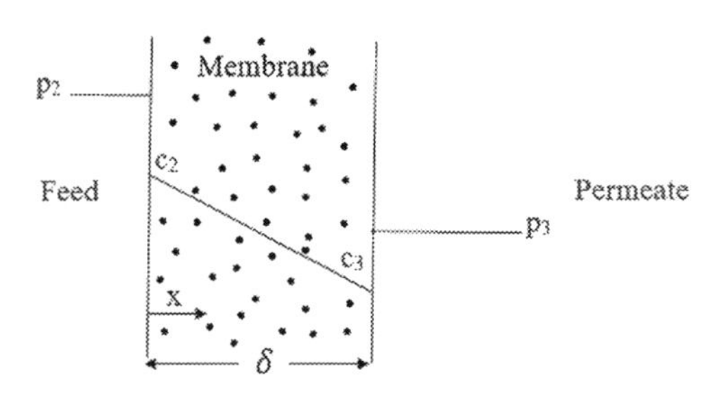

Figure 6.1 Solution-diffusion of gas transport across the membrane.

and P is the permeability. Its dimension is mol m/m^2 s Pa in the SI system. In the membrane literature, Barrer is often used, where 1 Barrer is equal to 10^{-10} cm^3(STP) cm/cm^2 s cmHg.

For the integrally asymmetric membrane, it is difficult to know the thickness of the selective layer precisely. In this case, P/δ, called permeance, is used instead of permeability. Its dimension is mol/m^2 s Pa in the SI system but gas permeation unit [1 GPU = 10^{-6} cm^3 (STP)/cm^2 s cmHg] is often used.

The temperature dependence of D follows the Arrhenius relationship,

$$D = D_0 \exp\left(-\frac{E_D}{RT}\right) \tag{6.6}$$

where D_0 and E_D are the pre-exponential factor and the activation energy of diffusion, respectively. The activation energy is considered as the energy required to create a gap between polymer segments through which the gas molecule can diffuse.

The solubility coefficient, S, also changes with temperature exponentially as

$$S = S_0 \exp\left(-\frac{\Delta H_s}{RT}\right) \tag{6.7}$$

where S_0 and ΔH_s are the pre-exponential factor and enthalpy of solution, respectively.

Combining Eqs. (6.6) and (6.7)

$$P = D_0 S_0 \exp\left(-\frac{E_D + \Delta H_s}{RT}\right) \tag{6.8}$$

$$= P_0 \exp\left(-\frac{E_p}{RT}\right) \tag{6.9}$$

where

$$E_p = E_D + \Delta H_s \tag{6.10}$$

It should be noted that E_p is the sum of E_D and ΔH_s. While E_D is always positive, ΔH_s may be positive or negative depending on whether the sorption is exothermic or endothermic.

Therefore, depending on their magnitude, E_p may be either positive or negative (Vieth, 1991).

The S-D model is a simple and general description of the gas transport. The manifestation of the fundamental principles of the transport depends on whether the membrane polymer is in a rubbery or glassy state. All polymers undergo transition from a rubbery to glassy state when the temperature is lowered below the glass transition temperature, T_g, which is characteristic to the polymer.

Problem 6.1:

Kurczek made a gas permeation experiment with a membrane made of sulfonated polyphenylene oxide (SPPO) (sulfonation degree 1.01 meq/g in hydrogen form) with a constant pressure system (Kruczek, 1999). The permeation rate of 0.471 cm^3/min was obtained at the temperature of 25°C and pressure of 763 cmHg.

The membrane area was 10.2 cm^2 and the transmembrane pressure difference was 517 cmHg. Calculate the permeance.

Answer:

The permeation rate at standard temperature and pressure (STP) is

$$0.471 \times \frac{273.15}{(273.15+25)} \times \frac{763}{760} \times \frac{1}{60} = 7.22 \times 10^{-3}\ \text{cm}^3(\text{STP})/\text{s}$$

Hence the permeance is

$$\frac{7.22 \times 10^{-3}}{(10.2 \times 517)} \times 10^{6} = 1.369\ \text{GPU}$$

The volume of 1 mole of ideal gas is 22.4×10^3 cm^3 (STP). Hence, the permeance in SI system is

$$1.369 \times 10^{-6}\text{cm}^3(\text{STP}) \times (\text{mol}/22.4 \times 10^3\text{cm}^3(\text{STP}))/(\text{cm}^2/(\text{cm}^2/10^{-4}\text{m}^2)\text{s}$$
$$(\text{cmHg}/(\text{cmHg}/1333.22\text{Pa}))$$

$$= \frac{1.369 \times 10^{-6}}{(22.4 \times 10^{-4} \times 1333.22)} = 4.584 \times 10^{-10}\text{mol}/\text{m}^2\text{s Pa}$$

6.1.2 Unsteady-state evaluation of S and D by the time-lag method

According to Fick's second law, gas diffusion in the membrane is given by

$$\frac{dc}{dt} = -D\frac{d^2c}{dx^2} \tag{6.11}$$

The differential Eq. (6.11) is solved under the following initial (IC) and boundary (BC) conditions,

$$\text{IC}: c(x, 0) = 0 \tag{6.12}$$

$$\text{BC1}: c(0, t) = Sp_2 \tag{6.13}$$

$$\text{BC2}: c(\delta, t) = 0 \tag{6.14}$$

To satisfy the initial and boundary conditions, the experiments are carried out by the device illustrated in Fig. 6.2 with the following protocol.

Step 1: The feed chamber, membrane, and permeate chamber are evacuated to satisfy IC.

Step 2: The pressure of the feed chamber is elevated to p_2 at $t = 0$.

Step 3: The pressure of the feed chamber is kept constant at p_2 throughout the experiment (BC1).

The pressure in the permeate chamber is kept at close to zero all the time (BC2), but it increases gradually in the closed permeate chamber (Flaconnèche et al., 2001).

Then, the gas concentration profile changes with time, as shown in Fig. 6.3, until the steady state is reached.

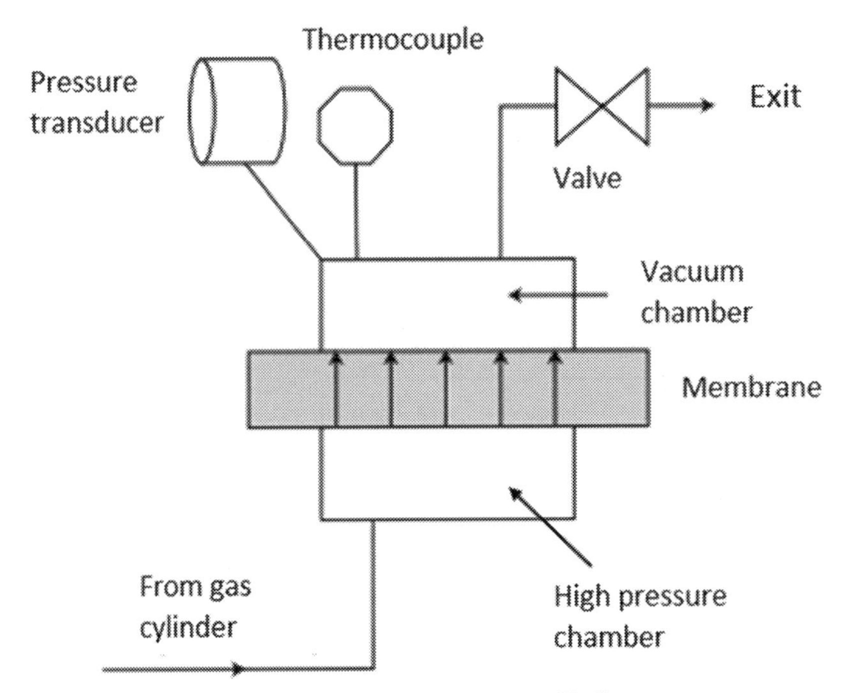

Figure 6.2 Device for the time lag method (Flaconnèche et al., 2001 modified).

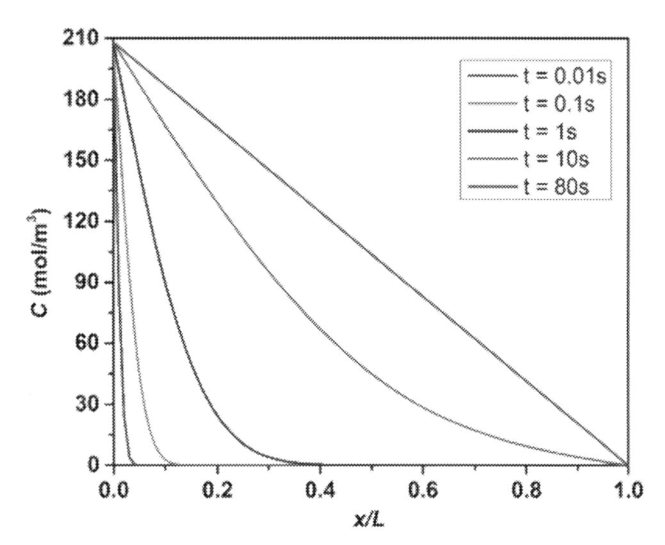

Figure 6.3 A typical example of concentration profile change in the membrane with time (Wu, 2020).

By solving the differential Eq. (6.11) under the initial and boundary conditions [Eqs. (6.12)–(6.14)], the concentration of gas in the membrane is given as a function of x and t by

$$c(x,t) = Sp_2\left(1-\frac{x}{\delta}\right) - \frac{2Sp_2}{\pi} \times \sum_{n=1}^{\infty}\frac{1}{n}\sin\left(\frac{n\pi x}{\delta}\right)\exp\left(-\frac{Dn^2\pi^2 t}{\delta^2}\right) \tag{6.15}$$

Applying Fick's first law Eq. (6.1) at $x=\delta$

$$J = J(\delta,t) = \frac{DSp_2}{\delta} + \frac{2DSp_2}{\delta} \times \sum_{n=1}^{\infty}\cos(n\pi)\exp\left(-\frac{Dn^2\pi^2 t}{\delta^2}\right) \tag{6.16}$$

Further by applying the ideal gas law, the pressure in the permeate chamber, p_3, is given as a function of t.

$$p_3(t) = p_3(\delta,t) = -\frac{ART}{V}\int_0^t J(\delta,t)dt$$

$$= \frac{ARTPp_2}{V\delta}\left\{t - \frac{\delta^2}{6D} + \frac{2\delta^2}{\pi^2 D} \times \sum_{n=1}^{\infty}\frac{(-1)^{n+1}}{n^2}\exp\left(-\frac{Dn^2\pi^2 t}{\delta^2}\right)\right\} \tag{6.17}$$

where A is the membrane area and V is the volume of the permeate chamber. R and T are the gas constant and absolute temperature, respectively.

For a long time t $\left(\frac{tD}{\delta^2} > 0.2\right)$, the exponential terms can be ignored and Eq. (6.17) can be approximated by a linear function

$$p_3(t) \cong \frac{ARTPp_2}{V\delta}\left(t - \frac{\delta^2}{6D}\right) \tag{6.18}$$

The typical permeate chamber pressure, $p_3(t)$ versus t is illustrated in Fig. 6.4, where the t axis intercept of the linear portion of the curve is the time lag θ.

Then, from Eq. (6.18), the following gas transport parameters can be obtained (Crank and Park, 1968):

$$D = \frac{\delta^2}{6\theta} \tag{6.19}$$

$$P = \text{Slope} \times \frac{V\delta}{ARTp_2} \tag{6.20}$$

and

$$S = \frac{P}{D} \tag{6.21}$$

Problem 6.2:

The gas permeation experiment was carried out using the device illustrated in Fig. 6.2 under the following conditions.

Membrane area $A = 23 \times 10^{-3}\ \text{m}^2$

Permeate chamber $V = 1.66 \times 10^{-4}\ \text{m}^3$

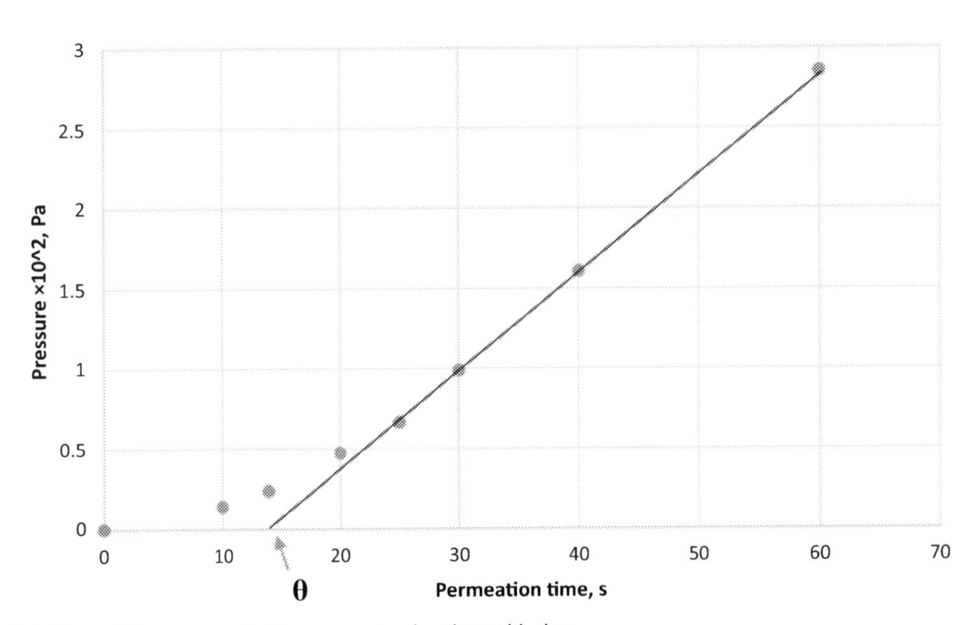

Figure 6.4 Rise of the pressure in the permeate chamber with time.

Table 6.1 Increase of pressure in the permeate chamber with time.

Time, t, s	Pressure in the permeate chamber, $p_3(t)$, Pa
0	0
10	14.2
14	23.9
20	47.6
25	66.7
30	99.0
40	160.9
60	285.7

Membrane thickness $\delta = 180 \times 10^{-6}$ m
Pressure of feed chamber $p_2 = 1.013 \times 10^5$ Pa
Temperature $T = 25°C$ and
the data given in Table 6.1 were obtained.

Calculate diffusivity, D, permeability, P, and solubility coefficient, S.

Answer:

From the above data the t axis intercept of the linear portion of the curve, θ, is 14 s (see Fig. 6.4).

Then, from Eq. (6.19)

$$D = \frac{\left(180 \times 10^{-6}\right)^2}{6 \times 14} = 3.86 \times 10^{-10} \text{m}^2/\text{s}$$

The slope is

$$\frac{285.7}{(60 - 14)} = 6.21 \text{ Pa/s}$$

Then, from Eq. (6.20)

$$P = 6.21 \times \frac{1.66 \times 10^{-5} \times 180 \times 10^{-6}}{23 \times 10^{-4} \times 8.314 \times 298.2 \times 1.013 \times 10^5}$$

$$= 3.21 \times 10^{-14} \frac{\text{mol m}}{\text{m}^2\text{s Pa}} = 95.8 \text{ Barrer}$$

Further, from Eq. (6.21)

$$S = \frac{P}{D} = \frac{3.21 \times 10^{-14}}{3.86 \times 10^{-10}} = 0.832 \times 10^{-4} \frac{\text{mol/m}^3}{\text{Pa}}$$

6.1.3 Separation of binary gas mixture

For a mixture of gas a and b, Eq. (6.5) is written as

$$J_i = P_i \frac{(p_{2,i} - p_{3,i})}{\delta} \quad i = a \text{ or } b \tag{6.22}$$

where J_i, P_i/δ, $p_{2,i}$, and $p_{3,i}$, are flux, permeance, and partial pressure in the feed chamber, and partial pressure in the permeate chamber of i-th gas, respectively.

The ratio of the permeances,

$$\frac{P_a}{P_b} = \frac{D_a}{D_b} \times \frac{S_a}{S_b} \tag{6.23}$$

is called selectivity of gas a over gas b. The ratios of diffusivity, D_a/D_b, and solubility coefficient, S_a/S_b, are called diffusivity selectivity and solubility selectivity, respectively.

$$\text{Therefore, selectivity} = \text{diffusivity selectivity} \times \text{solubility selectivity} \tag{6.24}$$

When the total pressure in the feed gas chamber is p_2 and the mole fractions of gas a and gas b are x_a and x_b, respectively, the mole fractions y_a and y_b in the permeate chamber of total pressure p_3 can be calculated as follows.

$$J_a = P_a(p_2 x_a - p_3 y_a) \tag{6.25}$$

$$J_b = P_b(p_2 x_b - p_3 y_b) \tag{6.26}$$

$$y_a = \frac{J_a}{J_a + J_b} = \frac{P_a(p_2 x_a - p_3 y_a)}{P_a(p_2 x_a - p_3 y_a) + P_b(p_2 x_b - p_3 y_b)} \tag{6.27}$$

$$y_a\{P_a(p_2 x_a - p_3 y_a) + P_b(p_2 x_b - p_3 y_b)\} = P_a(p_2 x_a - p_3 y_a) \tag{6.28}$$

$$y_a\{P_a(p_2 x_a - p_3 y_a) + P_b(p_2 x_b - p_3(1 - y_a))\} = P_a(p_2 x_a - p_3 y_a) \tag{6.29}$$

$$(P_b p_3 - P_a p_3)y_a^2 + (P_a p_2 x_a + P_b p_2 x_b + P_a p_3 - P_b p_3)y_a - P_a p_2 x_a = 0 \tag{6.30}$$

Solving the quadratic equation,

$$y_a = \frac{-(P_a p_2 x_a + P_b p_2 x_b + P_a p_3 - P_b p_3) + X}{2(P_b p_3 - P_a p_3)} \tag{6.31}$$

where

$$X = \sqrt{(P_a p_2 x_a + P_b p_2 x_b + P_a p_3 - P_b p_3)^2 + 4 \times (P_b p_3 - P_a p_3) \times P_a p_2 x_a} \tag{6.32}$$

$$\begin{aligned} y_b = 1 - y_a &= 1 - \frac{-(P_a p_2 x_a + P_b p_2 x_b + P_a p_3 - P_b p_3) + X}{2(P_b p_3 - P_a p_3)} \\ &= \frac{2(P_b p_3 - P_a p_3) + (P_a p_2 x_a + P_b p_2 x_b + P_a p_3 - P_b p_3) - X}{2(P_b p_3 - P_a p_3)} \\ &= \frac{-(-P_a p_2 x_a - P_b p_2 x_b + P_a p_3 - P_b p_3) - X}{2(P_b p_3 - P_a p_3)} \end{aligned} \tag{6.33}$$

Then,

$$\alpha = \frac{y_a/y_b}{x_a/x_b} = \frac{-(P_a p_2 x_a + P_b p_2 x_b + P_a p_3 - P_b p_3) + X}{-(-P_a p_2 x_a - P_b p_2 x_b + P_a p_3 - P_b p_3) - X} \times \frac{x_b}{x_a} \tag{6.34}$$

α is called separation factor.

Especially when $p_3 = 0$, from Eq. (6.27)

$$y_a = \frac{P_a p_2 x_a}{P_a p_2 x_a + P_b p_2 x_b} \tag{6.35}$$

$$y_b = \frac{P_b p_2 x_b}{P_a p_2 x_a + P_b p_2 x_b} \tag{6.36}$$

$$\alpha = \frac{y_a/y_b}{x_a/x_b} = \frac{(P_a p_2 x_a)/(P_b p_2 x_b)}{x_a/x_b} = P_a/P_b \tag{6.37}$$

Thus the separation factor is the same as selectivity when $p_3 = 0$. For $p_3 > 0$,

$\alpha < P_a/P_b$, as will be shown by the example.

Problem 6.3:

Kruczek et al. fabricated a membrane from SPPO in hydrogen form with an ionic content of 1.37 meq/g. Using a constant pressure gas permeation setup, they reported that the permeability of the membrane for CO_2 (P_{CO_2}) and methane (P_{CH_4}) was 16.86 Barrer and 0.41 Barrer, respectively. When the membrane is used for the separation of CO_2/CH_4 (0.5/0.5 mole fraction) mixture at the total feed pressure of 593 cmHg (absolute) and total permeate pressure of 76 cmHg (absolute, atmospheric pressure), what will be the separation factor?

Answer:
Inserting in Eq. (6.32)
$P_a = P_{CO2} = 16.86$ Barrer
$P_b = P_{CH4} = 0.41$ Barrer
$P_2 = 593$ cmHg
$P_3 = 76$ cmHg

$$X = \left((16.86 \times 593 \times 0.5 + 0.41 \times 593 \times 0.5 + 16.86 \times 76 - 0.41 \times 76)^2 + 4 \times (0.41 \times 76 - 16.81 \times 76) \times 16.86 \times 593 \times 0.5\right)^{\frac{1}{2}} = 3957.7$$

From Eq. (6.34),

$$\alpha = \frac{-(16.86 \times 593 \times 0.5 + 0.41 \times 593 \times 0.5 + 16.86 \times 76 - 0.41 \times 76) + 3957.7}{-(-16.86 \times 593 \times 0.5 - 0.41 \times 593 \times 0.5 + 16.86 \times 76 - 0.41 \times 76) - 3957.7} \times \frac{0.5}{0.5} = 27.6$$

On the other hand,

$$\frac{P_a}{P_b} = \frac{16.86}{0.41} = 41.12$$

Thus, the separation factor is much smaller than selectivity.

6.1.4 Resistance model

The multilayered membrane is defined as a membrane that consists of several barrier layers of distinct nature stacked together. There is a clear discontinuity at the boundary of two neighboring barrier layers, either in the chemical structure or in the morphology.

Henis and Tripodi derived an equation to evaluate the performance of the multilayered membrane based on the law of electrical circuit (Henis & Tripodi, 1977).

The permeation rate for a component i through a membrane can be written as

$$Q_i = \frac{P_i A \Delta p_i}{\delta} \quad (6.38)$$

where Q_i is the permeation rate of the component i, P_i is the permeability coefficient of the membrane, A is the surface area of the membrane, and δ is the thickness of the membrane. Considering an electric circuit, the following Ohm's law describes the current I, flowing through a resistance, R, driven by an electrical potential difference, E.

$$I = \frac{E}{R} \quad (6.39)$$

Further considering $Q_i = I$ and $\Delta p_i = E$,

$$R_i = \frac{\delta}{P_i A} \tag{6.40}$$

Accordingly, Eq. (6.38) can be written as

$$Q_i = \frac{\Delta p_i}{R_i} \tag{6.41}$$

The resistance model allows calculation of the overall resistance of a composite membrane as follows.

Case 1: Two resistances connected in series

When the barrier layers of resistances R_1 and R_2, respectively, are combined in series [see Fig. 6.5 (Matsuura, 1994)], layer 1 on top of layer 2, and two different gases, *a* and *b*, flow through the membrane, overall resistance for each permeant is given by

$$(R)_a = (R_1)_a + (R_2)_a \tag{6.42}$$

$$(R)_b = (R_1)_b + (R_2)_b \tag{6.43}$$

Defining the ratio of the resistances for the permeant *a* and *b* as

$$\alpha_i = \frac{(R_i)_b}{(R_i)_a} = \frac{(P_i)_a}{(P_i)_b} \tag{6.44}$$

where $i = 1$ and 2,
the ratio of the overall resistance for permeants *a* and *b* designated as α then becomes,

$$\begin{aligned} \alpha &= \frac{(R)_b}{(R)_a} \\ &= \frac{\alpha_1 (R_1)_a + \alpha_2 (R_2)_a}{(R_1)_a + (R_2)_a} \\ &= \frac{\alpha_1 + \alpha_2 \left[(R_2)_a/(R_1)_a\right]}{1 + \left[(R_2)_a/(R_1)_a\right]} \end{aligned} \tag{6.45}$$

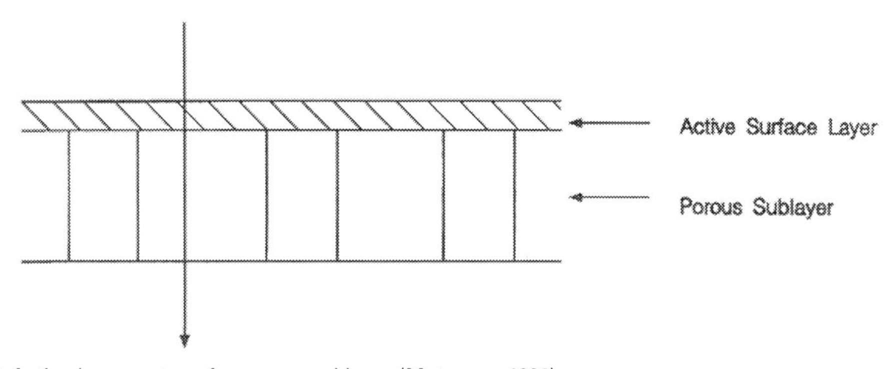

Figure 6.5 Active layer on top of a porous sublayer (Matsuura, 1994).

When $\left[(R_2)_a/(R_1)_a\right] \approx 0$

$$\alpha = \alpha_1 \tag{6.46}$$

The above equation indicates that the selectivity of the composite membrane is controlled by the barrier layer whose resistance is much higher than the other. This is the principle underlying the design of the composite membrane where a selective layer is supported by a porous substrate layer. The resistance of the former layer is much higher than the latter because of its nonporous characteristics.

Case 2: Two resistances connected in parallel

When barrier layers of resistance R_2 and R_3 are combined in parallel (see Fig. 6.6), the overall resistances for gases a and b are given by

$$R_a = \frac{(R_2)_a \times (R_3)_a}{(R_2)_a + (R_3)_a} \tag{6.47}$$

$$R_b = \alpha R_a = \frac{\alpha_2 (R_2)_a \times \alpha_3 (R_3)_a}{\alpha_2 (R_2)_a + \alpha_3 (R_3)_a} \tag{6.48}$$

Division of Eq. (6.48) by Eq. (6.47) yields

$$\alpha = \alpha_2\alpha_3 \frac{(R_2)_a \times (R_3)_a}{\alpha_2 (R_2)_a + \alpha_3 (R_3)_a} = \alpha_2\alpha_3 \frac{1 + \frac{(R_3)_a}{(R_2)_a}}{\alpha_2 + \alpha_3 \frac{(R_3)_a}{(R_2)_a}} \tag{6.49}$$

When R_3 is much smaller than R_2, $R_3/R_2 \cong 0$. Then,

$$\alpha \cong \alpha_3 \tag{6.50}$$

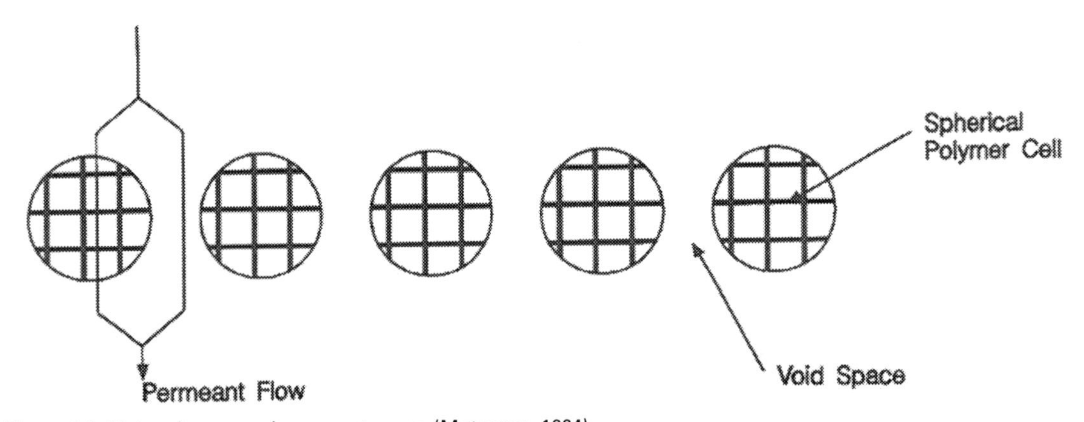

Figure 6.6 Network pore and aggregate pore (Matsuura, 1994).

Therefore the overall selectivity is controlled by the component with a smaller resistance.

According to the bimodal pore size distribution, there are two kinds of pores; one is the network pore whose size is smaller, and the other is the aggregate pore whose size is much larger than the network pore. Since the resistance of the aggregate pore is much smaller, the overall selectivity is controlled by the aggregate pore of lower selectivity. However, as the number of aggregate pores becomes smaller, the total area of the aggregate pores becomes smaller and the network pores start to control the selectivity. Therefore, the aggregate pores have to be removed as much as possible to improve the overall selectivity.

Case 3: Two series resistances connected in parallel

When two arms of resistance, $R_1 + R_2$ and $R_1' + R_3$ are connected in parallel (see Fig. 6.7), the overall resistances for gas a and b are

$$R_a = \frac{[(R_1)_a + (R_2)_a] \times \left[(R_1')_a + (R_3)_a\right]}{\left[(R_1)_a + (R_2)_a + (R_1')_a + (R_3)_a\right]} \tag{6.51}$$

and

$$R_b = \alpha R_a = \frac{[\alpha_1(R_1)_a + \alpha_2(R_2)_a] \times \left[\alpha_1'(R_1')_a + \alpha_3(R_3)_a\right]}{\left[\alpha_1(R_1)_a + \alpha_2(R_2)_a + \alpha_1'(R_1')_a + \alpha_3(R_3)_a\right]} \tag{6.52}$$

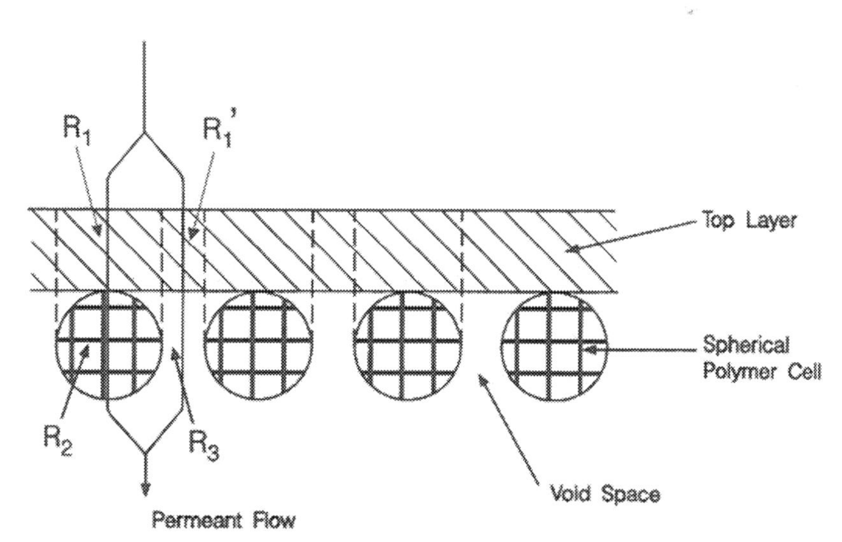

Figure 6.7 Homogeneous membrane laminated on top of a porous substrate membrane (Matsuura, 1994).

When

$$R_1 + R_2 \ll R_1^{'} + R_3$$

And also

$$R_2 \gg R_1$$

R_2 governs the overall resistance.

Case 4: Wheatstone bridge model

When the two arms of resistance, $R_1^{'} + R_2$ and $R_1 + R_3$ are connected in parallel, and also by a cross-flow resistance R_x (see Fig. 6.8), a Wheatstone bridge is formed. Then the overall resistance becomes

$$R = \frac{R_x \times (R_1 + R_2)(R_1^{'} + R_3) + R_1 R_1^{'}(R_2 + R_3) + R_2 R_3 (R_1 + R_1^{'})}{R_x \times (R_1 + R_2 + R_1^{'} + R_3) + (R_1 + R_1^{'})(R_2 + R_3)} \tag{6.53}$$

When $R_x = \infty$

Eq. (6.53) for gas *a* becomes the same as Eq. (6.51).

When $R_x = 0$

$$R = \frac{R_1 \times R_1^{'}}{(R_1 + R_1^{'})} + \frac{R_2 \times R_3}{(R_2 + R_3)} \tag{6.54}$$

In the first term of the left-hand side, if $R_1^{'} \gg R_1$ (e.g., area of resistance $R_1^{'}$ is much smaller than R_1)

Eq. (6.54) becomes

$$R = R_1 + \frac{R_2 \times R_3}{(R_2 + R_3)} \tag{6.55}$$

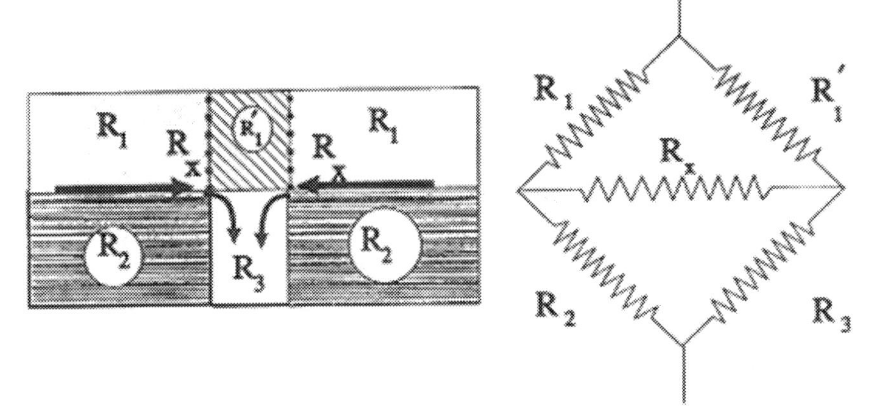

Figure 6.8 Wheatstone bridge model (Matsuura, 1994).

Eq. (6.55) is Henis-Tripodi's model. In their prism membrane (see Fig. 6.9), a polysulfone membrane with defective pores is coated with a layer of silicone rubber by dip coating and during the dip coating process the pores are filled with silicone rubber.

Then, R_1, R_2 and R_3, respectively, represent the resistance of the top silicone rubber layer, the polysulfone matrix, and the pores filled with silicone rubber.

Since the area of silicone rubber-filled pores is much smaller than that of the polysulfone matrix,

$$R_3 \gg R_2$$

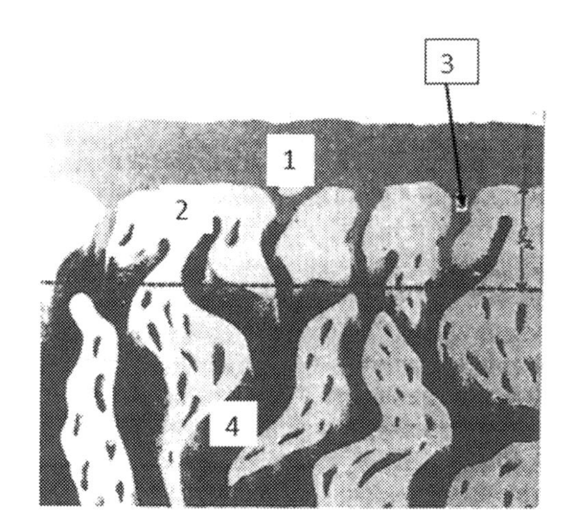

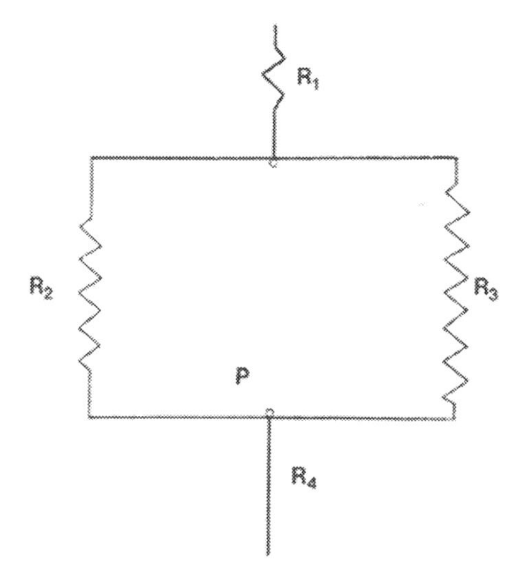

Figure 6.9 Henis–Tripodi model (Matsuura, 1994).

And since permeability of silicone rubber is much larger than that of polysulfone,

$$R_2 \gg R_1$$

Then, from Eq. (6.55), $R \approx R_2$.

Problem 6.4:

There is a polysulfone porous substrate with an area ratio (pore/polymer matrix) $A_3/A_2 = 1.9 \times 10^{-6}$ and the effective film thickness 10^{-7} m (100 nm). The permeability of H_2 and CO for polysulfone is 4.019×10^{-13} mol m/m^2 s Pa (1200×10^{-10} cc [STP] cm/cm^2 s cmHg) and 10.047×10^{-15} mol m/m^2 s Pa (30×10^{-10} cc[STP] cm/cm^2 s cmHg), respectively. The substrate is coated with silicone rubber with permeabilities of 17.415×10^{-12} mol m/m^2 s Pa (520×10^{-8} cc[STP] cm/cm^2 s cmHg) and 8.373×10^{-12} mol m/m^2 s Pa (250×10^{-8} cc[STP] cm/cm^2 s cmHg) for H_2 and CO, respectively. The coating thickness is 10^{-6} m (1 μm). What is the overall resistance of the composite membrane for H_2 and CO?

Answer:

$$(R_1)_{H_2} = \frac{(10^{-6})}{(17.415 \times 10^{-12})A_1}$$

$$= \frac{0.05742 \times 10^6}{A_1}$$

$$(R_1)_{CO} = \frac{(10^{-6})}{(8.373 \times 10^{-12})A_1}$$

$$= \frac{0.1194 \times 10^6}{A_1}$$

$$(R_2)_{H_2} = \frac{(10^{-7})}{(4.019 \times 10^{-13})A_2}$$

$$= \frac{0.2488 \times 10^6}{A_2}$$

$$(R_2)_{CO} = \frac{(10^{-7})}{(10.047 \times 10^{-15})A_2}$$

$$= \frac{0.09953 \times 10^8}{A_2}$$

$$(R_3)_{H_2} = \frac{(10^{-7})}{(17.415 \times 10^{-12})A_3}$$

$$= \frac{0.05742 \times 10^5}{A_3}$$

$$(R_3)_{CO} = \frac{(10^{-7})}{(8.373 \times 10^{-12})A_3}$$

$$= \frac{0.1194 \times 10^5}{A_3}$$

From Eq. (6.55)

$$(R)_{H_2} = 0.05742 \times 10^6/A_1 + \frac{(0.2488 \times 10^6/A_2) \times (0.05742 \times 10^5/A_3)}{(0.2488 \times 10^6/A_2) + (0.05742 \times 10^5/A_3)}$$

$$(R)_{CO} = 0.1194 \times 10^6/A_1 + \frac{(0.09953 \times 10^8/A_2) \times (0.1194 \times 10^5/A_3)}{(0.09953 \times 10^8/A_2) + (0.1194 \times 10^5/A_3)}$$

Assuming $A_1 = A_2$, since $A_3 = 1.9 \times 10^{-6} A_2$, the permeability ratio α = (permeability of H_2/permeability of CO) becomes

$$\frac{1}{\alpha} = \frac{0.05742 \times 10^6 + \frac{(0.2488 \times 10^6) \times (0.05742 \times 10^5/1.9 \times 10^{-6})}{(0.2488 \times 10^6) + (0.05742 \times 105/1.9 \times 10^{-6})}}{0.1194 \times 10^6/A_1 + \frac{(0.09953 \times 10^8) \times (0.1194 \times 10^5/1.9 \times 10^{-6})}{(0.09953 \times 10^8) + (0.1194 \times 10^5/1.9 \times 10^{-6})}} = 0.03041$$

$$\alpha = 33$$

The H_2/CO permeability ratio of polysulfone is $4.019 \times 10^{-13}/10.047 \times 10^{-15} = 40$. Thus, the selectivity of the layered membrane is only less than 20% lower than that of polysulfone.

Applying the two parallel resistance models, the resistance of the substrate membrane, whose pores are filled with silicone rubber without the top silicone rubber coating, is

$$(R_{sub})_{H_2} = \frac{(0.2488 \times 10^6/A_2) \times (0.05742 \times 10^5/A_3)}{(0.2488 \times 10^6/A_2) + (0.05742 \times 105/A_3)}$$

The ratio of the flux of the coated membrane to that of the substrate membrane is

$$\frac{(R_{sub})_{H_2}}{(R)_{H_2}} = 0.8125$$

About 20% of the flux was sacrificed by the silicone coating.

Problem 6.5:

A multilayered membrane was made by laminating a silicone rubber film on top of a porous polysulfone substrate membrane. The following data were obtained by experiments for the silicone rubber film, porous substrate, and the multilayered membrane.

$(P_1/\delta)_{H_2}$: Hydrogen permeance of silicon rubber film, 80.18×10^{-10} mol/m^2 sPa

$(P_{sub}/\delta)_{H_2}$: Hydrogen permeance of substrate membrane, 112.2×10^{-10} mol/m^2 s Pa

$(P/\delta)_{H_2}$: Hydrogen permeance of the multilayered membrane, 29.9×10^{-10} mol/m^2 s Pa

α_1: H_2/N_2 permeability ratio of silicone rubber film, 2.2

α_{sub}: H_2/N_2 permeability ratio of the substrate membrane, 6.29

α: H_2/N_2 permeability ratio of the multilayered membrane, 36.9

It is known that the H_2/N_2 permeability ratio of polysulfone polymer matrix, α_2, is 66.67

The total membrane area is 10.18×10^{-4} m^2. Calculate the resistance of each component.

Answer:

Using Eq. (6.40)

$$(R_1)_{H_2} = \frac{1}{(80.18 \times 10^{-10}) \times (10.18 \times 10^{-4})} = 12.19 \times 10^{10} s\ \text{Pa/mol}$$

$$(R_{sub})_{H_2} = \frac{1}{(112.2 \times 10^{-10}) \times (10.18 \times 10^{-4})} = 8.755 \times 10^{10} s\ \text{Pa/mol}$$

$$(R)_{H_2} = \frac{1}{(29.2 \times 10^{-10}) \times (10.18 \times 10^{-4})} = 33.64 \times 10^{10} s\ \text{Pa/mol}$$

In case 3 (two series resistances connected in parallel) the following symbols for the unknown resistances $x = (R_2)_{H_2}$, $y = (R_3)_{H_2}$, and $z = (R_1')_{H_2}$ are used.

First, applying Eqs. (6.47) and (6.48) of case 2 (two resistances connected in parallel) for the substrate membrane,

$$8.755 \times 10^{10} = \frac{xy}{x+y} \tag{6.56}$$

$$6.29 \times 8.755 \times 10^{10} = \frac{(66.67x) \times (\alpha_3 y)}{(66.67x) + (\alpha_3 y)} \tag{6.57}$$

Then, applying Eqs. (6.51) and (6.52) of case 3 (two series resistances connected in parallel),

$$33.64 \times 10^{10} = \frac{\left[\left(12.19 \times 10^{10}\right) + x\right] \times (z + y)}{\left(12.19 \times 10^{10}\right) + x + (z + y)} \tag{6.58}$$

$$36.9 \times 33.64 \times 10^{10} = \frac{\left[\left(2.2 \times 12.19 \times 10^{10}\right) + 66.67x\right] + \left(2.2z + \alpha_3 y\right)}{\left(2.2 \times 12.19 \times 10^{10}\right) + 66.67x + \left(2.2z + \alpha_3 y\right)} \tag{6.59}$$

Division of Eqs. (6.57) by (6.56) yields

$$6.29 = 66.67\alpha_3 \frac{x + y}{(66.67x) + \left(\alpha_3 y\right)} \tag{6.60}$$

For Eqs. (6.56) and (6.58)–(6.60), there are four unknowns, x, y, z, and α_3, which can be solved by using the algorithm given in Scheme 6.1.

The results are:

$x = 21.78 \times 10^{10}$, $y = 14.64 \times 10^{10}$, $z = 3448.2 \times 10^{10}$ and $\alpha_3 = 3.91$.

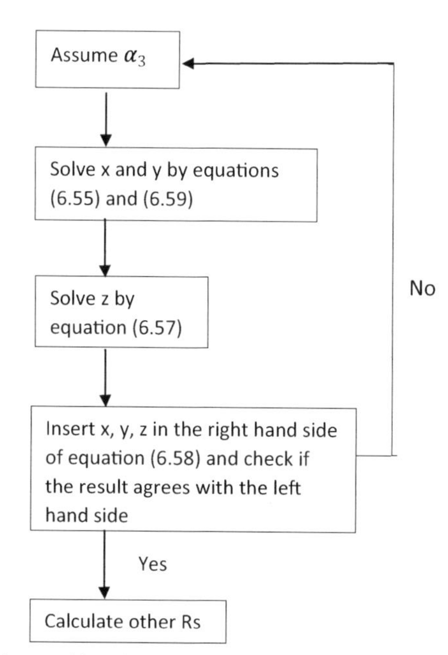

Scheme 6.1 Algorithm used to solve problem 6.5.

and, the resistances are:

$$(R_2)_{H_2} = 21.78 \times 10^{10} s\ \mathrm{Pa/mol}$$

$$(R_3)_{H_2} = 14.64 \times 10^{10} s\ \mathrm{Pa/mol}$$

$$\left(R_1'\right)_{H_2} = 3448.2 \times 10^{10} s\ \mathrm{Pa/mol}$$

$$(R_2)_{N_2} = 1452 \times 10^{10} s\ \mathrm{Pa/mol}$$

$$(R_3)_{N_2} = 57.24 \times 10^{10} s\ \mathrm{Pa/mol}$$

$$\left(R_1'\right)_{N_2} = 7586 \times 10^{10} s\ \mathrm{Pa/mol}$$

6.2 Gas transport in porous membrane

6.2.1 Transport mechanism

Transport of gas can also take place through the porous membrane, even though the separation of gas mixtures is not as effective as by transport through nonporous membranes. Transport equations of porous membranes are often used to analyze the data obtained from membrane distillation experiments or to characterize membranes by pore size and pore size distribution.

As shown by Eq. (6.61) the flux of i-th species in the membrane is given by

$$J_i = P_i \frac{(p_{2,i} - p_{3,i})}{\delta} \quad i = a \text{ or } b \tag{6.61}$$

which is also applicable for the porous membrane.

Gas transport through the porous membrane is classified into (1) Knudsen flow, (2) viscous flow, and (3) the combination of Knudsen and viscous flow, depending on the gas pressure and the membrane pore size (Khayet and Matsuura, 2011). The governing quantity that provides a guideline in determining which mechanism is dominant is the Knudsen number, Kn, which is given by

$$Kn = \frac{\lambda}{d_p} \tag{6.62}$$

where λ is the mean free path of the gas molecules and d_p is the pore diameter. λ is further given by

$$\lambda = \frac{kT}{\sqrt{2}\pi\sigma^2 p} \tag{6.63}$$

where k is the Boltzmann constant, T is absolute temperature, σ is the collision diameter of the gas molecule, and p is pressure.

Knudsen flow

When the mean free path of the molecule is far greater than the membrane pore size, for example, $Kn > 10$, the molecule–pore wall collisions occur more frequently than the molecule–molecule collisions, as schematically illustrated in Fig. 6.10A and the Knudsen type flow becomes dominant. The gas permeance is given by the following equation.

$$\frac{P_i}{\delta} = \frac{2\varepsilon}{3}\frac{1}{RT}\left(\frac{8RT}{\pi M_i}\right)^{1/2}\frac{r}{\tau\delta} \tag{6.64}$$

where ε is membrane porosity, R is the universal gas constant, M is the molecular weight of gas, r is pore radius, τ is tortuosity factor, and δ is the membrane thickness.

Molecular diffusion

When the mean free path of the molecule is far smaller than the membrane pore size, say for example, $Kn < 0.01$, the molecule–molecule collisions occur more frequently than the molecule–pore wall collisions, as schematically illustrated in Fig. 6.10B and the molecular diffusion becomes dominant. The gas permeability is given by the following equation.

$$\frac{P_i}{\delta} = \frac{\varepsilon r^2}{8\eta_i}\frac{\overline{p}}{RT}\frac{1}{\tau\delta} \tag{6.65}$$

where η_i is the viscosity of gas and $\overline{p}$ is the average pressure in the pore.

Intermediate region

When $0.01 < Kn < 100$ both molecule–pore collisions and molecule–molecule collisions should be taken into consideration, and the gas permeability is given by the following equation.

$$\frac{P_i}{\delta} = \frac{\varepsilon}{RT\tau\delta}\left[\frac{2}{3}\left(\frac{8RT}{\pi M_i}\right)^{1/2} r + \frac{r^2}{8\eta_i}\overline{p}\right] \tag{6.66}$$

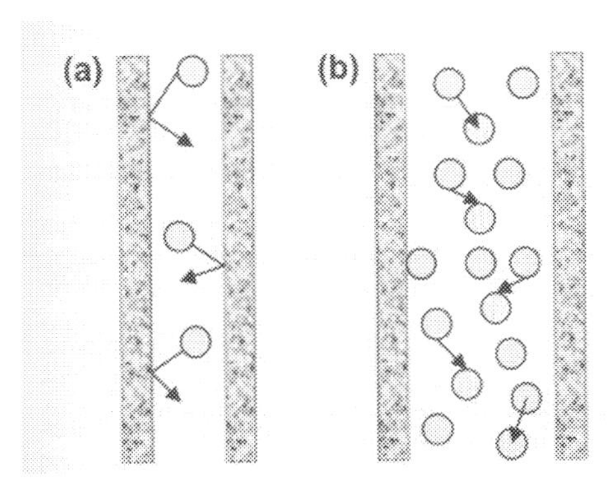

Figure 6.10 (A) Knudsen diffusion and (B) molecular diffusion (Khayet and Matsuura, 2011).

For pure gas transport, the application of the above equations is straightforward. For the gas mixture, the viscosity of the gas mixture is used for η_i in Eqs. (6.65) and (6.66).

6.2.2 Separation of gas mixture by the porous membrane

Knudsen flow

From Eqs. (6.56) and (6.64)

$$J_a = \frac{2\varepsilon}{3}\frac{1}{RT}\left(\frac{8RT}{\pi M_a}\right)^{\frac{1}{2}}\frac{r}{\tau\delta}(p_{2,a} - p_{3,a}) = \frac{2\varepsilon}{3}\frac{1}{RT}\left(\frac{8RT}{\pi M_a}\right)^{\frac{1}{2}}\frac{r}{\tau\delta}(p_2X_{2,a} - p_3X_{3,a}) \tag{6.67}$$

where X is mole fraction.

Similarly,

$$J_b = \frac{2\varepsilon}{3}\frac{1}{RT}\left(\frac{8RT}{\pi M_b}\right)^{\frac{1}{2}}\frac{r}{\tau\delta}(p_2X_{2,b} - p_3X_{3,b}) \tag{6.68}$$

Hence,

$$X_{3,a} = \frac{J_a}{J_a + J_b}$$

$$= \frac{\frac{2\varepsilon}{3}\frac{1}{RT}\left(\frac{8RT}{\pi M_a}\right)^{\frac{1}{2}}\frac{r}{\tau\delta}(p_2X_{2,a} - p_3X_{3,a})}{\frac{2\varepsilon}{3}\frac{1}{RT}\left(\frac{8RT}{\pi M_a}\right)^{\frac{1}{2}}\frac{r}{\tau\delta}(p_2X_{2,a} - p_3X_{3,a}) + \frac{2\varepsilon}{3}\frac{1}{RT}\left(\frac{8RT}{\pi M_b}\right)^{\frac{1}{2}}\frac{r}{\tau\delta}(p_2X_{2,b} - p_3X_{3,b})} \tag{6.69}$$

When $p_3 = 0$

$$X_{3,a} = \frac{\left(\frac{1}{M_a}\right)^{\frac{1}{2}}X_{2,a}}{\left(\frac{1}{M_a}\right)^{\frac{1}{2}}X_{2,a} + \left(\frac{1}{M_b}\right)^{\frac{1}{2}}X_{2,b}} \tag{6.70}$$

Similarly,

$$X_{3,b} = \frac{\left(\frac{1}{M_b}\right)^{\frac{1}{2}}X_{2,b}}{\left(\frac{1}{M_a}\right)^{\frac{1}{2}}X_{2,a} + \left(\frac{1}{M_b}\right)^{\frac{1}{2}}X_{2,b}} \tag{6.71}$$

Hence, the selectivity (ideal separation factor)

$$\alpha = \frac{X_{3,a}/X_{3,b}}{X_{2,a}/X_{2,b}} = \left(\frac{M_b}{M_a}\right)^{1/2} \tag{6.72}$$

The selectivity is reciprocal to the square root of molecular weight ratio.

Viscous flow

From Eqs. (6.61) and (6.65)

$$J_a = \frac{\varepsilon r^2}{8\eta_{mixt}} \frac{\overline{p}}{RT} \frac{1}{\tau\delta}\left(p_{2,a} - p_{3,a}\right) = \frac{\varepsilon r^2}{8\eta_{mixt}} \frac{\overline{p}}{RT} \frac{1}{\tau\delta}\left(p_2 X_{2,a} - p_3 X_{3,a}\right) \tag{6.73}$$

Similarly,

$$J_b = \frac{\varepsilon r^2}{8\eta_{mixt}} \frac{\overline{p}}{RT} \frac{1}{\tau\delta}\left(p_2 X_{2,b} - p_3 X_{3,b}\right) \tag{6.74}$$

$$X_{3,a} = \frac{J_a}{J_a + J_b} = \frac{\left(p_2 X_{2,a} - p_3 X_{3,a}\right)}{\left(p_2 X_{2,a} - p_3 X_{3,a}\right) + \left(p_2 X_{2,b} - p_3 X_{3,b}\right)} = \frac{\left(p_2 X_{2,a} - p_3 X_{3,a}\right)}{\left(p_2 - p_3\right)}$$

Hence

$$X_{3,a}\left(p_2 - p_3\right) = \left(p_2 X_{2,a} - p_3 X_{3,a}\right)$$

$$X_{3,a} = X_{2,a} \tag{6.75}$$

No separation takes place for any p_3.

Problem 6.6:

The collision diameter of nitrogen gas is 0.364 nm. Calculate the mean free path at the temperature of 25°C and pressures of 0.1, 0.5, and 1.0 bar. The Boltzmann constant is 1.381×10^{-23} J/K.

Answer:

According to Eq. (6.63)

$$\lambda = \frac{1.381 \times 10^{-23} \times 298.2}{1.414 \times 3.1416 \times \left(0.364 \times 10^{-9}\right)^2 \times 10^4} = 0.7 \times 10^{-6}\text{m at 0.1 bar}$$

$$= 0.14 \times 10^{-6}\text{m} \quad \text{at 0.5 bar}$$

$$= 0.07 \times 10^{-6}\text{m} \quad \text{at 1.0 bar}$$

Problem 6.7:

Calculate the Knudsen number and the permeance of nitrogen gas for a membrane of porosity 0.5, pore radii 10, 100, and 1000 nm, and pore length 1 μm. The average pressure in the pore and the temperature are 0.1 bar and 25°C, respectively. Use nitrogen gas viscosity of 1.77×10^{-5} Pa s and assume $\tau = 1$.

Answer:

$Kn = \frac{0.7 \times 10^{-6}}{20 \times 10^{-9}} = 35$ for the pore of 10 nm

The gas transport is in the Knudsen region. According to Eq. (6.64)

$$\frac{P_{N_2}}{\delta} = \frac{2\times 0.5}{3}\frac{1}{8.314\times 298.2}\left(\frac{8\times 8.314\times 298.2}{3.1416\times 28.02\times 10^{-3}}\right)^{1/2}\frac{10\times 10^{-9}}{10^{-6}}$$

$$= 0.640\times 10^{-3}\text{mol/m}^2\text{s Pa}$$

$Kn = \frac{0.7\times 10^{-6}}{200\times 10^{-9}} = 3.5$ for the pore of 100 nm

The gas transport is in the intermediate region. According to Eq. (6.66)

$$\frac{P_{N_2}}{\delta} = \frac{0.5}{8.314\times 298.2\times 10^{-6}}\left[\frac{2}{3}\left(\frac{8\times 8.314\times 298.2}{3.1416\times 28.02\times 10^{-3}}\right)^{\frac{1}{2}}\times 100\times 10^{-9}\right.$$

$$\left. + \frac{(100\times 10^{-9})^2}{8\times 1.77\times 10^{-5}}\times 10^4\right] = 0.653\times 10^{-2}\text{mol/m}^2\text{s Pa}$$

$Kn = \frac{0.7\times 10^{-6}}{2000\times 10^{-9}} = 0.35$ for the pore of 1000 nm

The gas transport is in the intermediate region. According to Eq. (6.66)

$$\frac{P_{N_2}}{\delta} = \frac{0.5}{8.314\times 298.2\times 10^{-6}}\left[\frac{2}{3}\left(\frac{8\times 8.314\times 298.2}{3.1416\times 28.02\times 10^{-3}}\right)^{\frac{1}{2}}\times 1000\times 10^{-9}\right.$$

$$\left. + \frac{(1000\times 10^{-9})^2}{8\times 1.77\times 10^{-5}}\times 10^4\right] = 0.780\times 10^{-1}\text{mol/m}^2\text{s Pa}$$

What about when the pressure is 1 bar?

$Kn = \frac{0.07\times 10^{-6}}{20\times 10^{-9}} = 3.5$ for the pore of 10 nm

The gas transport is in the intermediate region. According to Eq. (6.66)

$$\frac{P_{N_2}}{\delta} = \frac{0.5}{8.314\times 298.2\times 10^{-6}}\left[\frac{2}{3}\left(\frac{8\times 8.314\times 298.2}{3.1416\times 28.02\times 10^{-3}}\right)^{\frac{1}{2}}\times 10\times 10^{-9}\right.$$

$$\left. + \frac{(10\times 10^{-9})^2}{8\times 1.77\times 10^{-5}}\times 10^5\right] = 0.653\times 10^{-3}\text{mol/m}^2\text{s Pa}$$

$Kn = \frac{0.07\times 10^{-6}}{200\times 10^{-9}} = 0.35$ for the pore of 100 nm

The gas transport is in the intermediate region. According to Eq. (6.66)

$$\frac{P_{N_2}}{\delta} = \frac{0.5}{8.314 \times 298.2 \times 10^{-6}} \left[\frac{2}{3} \left(\frac{8 \times 8.314 \times 298.2}{3.1416 \times 28.02 \times 10^{-3}} \right)^{\frac{1}{2}} \times 100 \times 10^{-9} \right.$$

$$\left. + \frac{(100 \times 10^{-9})^2}{8 \times 1.77 \times 10^{-5}} \times 10^5 \right] = 0.780 \times 10^{-2} \text{mol/m}^2\text{s Pa}$$

$Kn = \frac{0.07 \times 10^{-6}}{2000 \times 10^{-9}} = 0.035$ for the pore of 1000 nm

The gas transport is in the intermediate region. According to Eq. (6.66)

$$\frac{P_{N_2}}{\delta} = \frac{0.5}{8.314 \times 298.2 \times 10^{-6}} \left[\frac{2}{3} \left(\frac{8 \times 8.314 \times 298.2}{3.1416 \times 28.02 \times 10^{-3}} \right)^{\frac{1}{2}} \times 1000 \times 10^{-9} \right.$$

$$\left. + \frac{(1000 \times 10^{-9})^2}{8 \times 1.77 \times 10^{-5}} \times 10^5 \right] = 2.063 \times 10^{-1} \text{mol/m}^2\text{s Pa}$$

Problem 6.8:

Calculate the flux and separation factor of air by a porous membrane with a porosity of 0.5, pore radius of 100 nm, and length of 1 μm at 25°C. The feed pressure is 5 bar (absolute) and the permeate is 1 bar (absolute). Viscosity of air at 25°C is equal to 1.86×10^{-5} Pa s. Use $\tau = 1$, and the oxygen mole fraction of 0.21 and nitrogen mole fraction of 0.79 for approximation.

Answer:

The collision diameters of nitrogen and oxygen are 0.364 and 0.346×10^{-9} m, respectively, which are close to each other. Therefore the average of 0.355×10^{-9} m is used in Eq. (6.63).

$$\lambda = \frac{kT}{\sqrt{2}\pi\sigma^2 p} = \frac{1.381 \times 10^{-23} \times 298.2}{1.414 \times 3.1416 \times (0.355 \times 10^{-9})^2 \times (3.0 \times 10^5)}$$

$$= 0.0245 \times 10^{-6} \text{m}$$

$$Kn = \frac{0.0245 \times 10^{-6}}{200 \times 10^{-9}} = 0.123$$

This is an intermediate region.

Setting a is oxygen and b is nitrogen,

$$J_a = \frac{\varepsilon}{RT\tau\delta}\left[\frac{2}{3}\left(\frac{8RT}{\pi M_a}\right)^{1/2} r + \frac{r^2}{8\eta_{mixt}}\overline{p}\right](p_2X_{2,a} - p_3X_{3,a})$$

$$= \left(\frac{0.5}{8.314\times 298.2\times 10^{-6}}\left[\frac{2}{3}\left(\frac{8\times 8.314\times 298.2}{3.1416\times 32.00\times 10^{-3}}\right)^{\frac{1}{2}}\times 100\times 10^{-9} + \frac{(100\times 10^{-9})^2}{8\times 1.86\times 10^{-5}}\times 3.0\times 10^5\right]\right)(5\times 10^5\times 0.21 - 1\times 10^5\times X_{3,a})$$

$$= 0.847\times 10^3 \tag{6.76}$$

$$J_b = \frac{\varepsilon}{RT\tau\delta}\left[\frac{2}{3}\left(\frac{8RT}{\pi M_b}\right)^{1/2} r + \frac{r^2}{8\eta_{\text{mixt}}}\overline{p}\right](p_2X_{2,b} - p_3X_{3,b})$$

$$= \left(\frac{0.5}{8.314\times 298.2\times 10^{-6}}\left[\frac{2}{3}\left(\frac{8\times 8.314\times 298.2}{3.1416\times 28.02\times 10^{-3}}\right)^{\frac{1}{2}}\times 100\times 10^{-9} + \frac{(100\times 10^{-9})^2}{8\times 1.86\times 10^{-5}}\times 3.0\times 10^5\right]\right)$$
$$(5\times 10^5\times 0.79 - 1\times 10^5\times (1 - X_{3,a})) = 3.310\times 10^3 \tag{6.77}$$

$$X_{3,b} = \frac{J_a}{J_a + J_b} \tag{6.78}$$

$X_{3,a} = 0.204$ and $X_{3,b} = 0.796$ satisfies Eqs. (6.76)–(6.78). There is barely any separation.

The total gas flux is

$$J_a + J_b = 0.847\times 10^3 + 3.310\times 10^3 = 4.157\times 10^3\ \text{mol/m}^2\ \text{s}$$

6.2.3 Measurement of pore size and pore size distribution

Measurement of pore size by the gas permeation method is described in K. Li's book (Li, 2007) in detail.

This is one of the simplest methods for measuring the membrane pore size. By measuring gas flow rates of a membrane at different pressures, the average pore size and effective surface porosity can be obtained.

According to Tan et al. (2001), and also as shown in Section 6.2.2, the flux through the membrane is written as the sum of the Poiseuille (viscous) and Knudsen flow as follows

$$J = J_v + J_k \tag{6.79}$$

Assuming that the pore is round and cylindrical with radius r_p and the number of the pore at the membrane surface is n per unit area of the membrane, porosity ε is

$$\varepsilon = n\pi {r_p}^2 \tag{6.80}$$

Then, the fluxes due to the Poiseuille and Knudsen flow are written, respectively (Present, 1958), as

$$J_v = \frac{\varepsilon {r_p}^2}{8\eta RT}\overline{p}\frac{\Delta p}{\delta} \tag{6.81}$$

$$J_k = \frac{2\varepsilon}{3}\left(\frac{8RT}{\pi M}\right)^{1/2}\frac{r_p}{RT}\frac{\Delta p}{\delta} \tag{6.82}$$

where η is the viscosity of the gas, R is the gas constant, T is the absolute temperature, $\overline{p}$ is the average pressure in the pore $\left(= \frac{p_2 + p_3}{2}\right)$, δ is the length of the pore, and M is the molecular weight of the gas.

From Eqs. (6.79), (6.81), and (6.82)

$$J = \frac{\varepsilon {r_p}^2}{8\eta RT}\overline{p}\frac{\Delta p}{\delta} + \frac{2\varepsilon}{3}\left(\frac{8RT}{\pi M}\right)^{1/2}\frac{r_p}{RT}\frac{\Delta p}{\delta} \tag{6.83}$$

Since δ is difficult to know, especially for asymmetric membranes, the flux is also given by the following equation

$$J = \frac{P}{\delta}\Delta p \tag{6.84}$$

where $\frac{P}{\delta}$ is permeance.

Then,

$$\frac{P}{\delta} = \frac{\varepsilon {r_p}^2}{8\eta RT\delta}\overline{p} + \frac{2\varepsilon}{3}\left(\frac{8RT}{\pi M}\right)^{1/2}\frac{r_p}{RT\delta} \tag{6.85}$$

When Eq. (6.85) is written as

$$\frac{P}{\delta} = P_0\overline{p} + K_0 \tag{6.86}$$

$$P_0 = \frac{\varepsilon {r_p}^2}{8\eta RT\delta} \tag{6.87}$$

$$K_0 = \frac{2\varepsilon}{3}\left(\frac{8RT}{\pi M}\right)^{1/2}\frac{r_p}{RT\delta} \tag{6.88}$$

Hence,

$$r_p = \frac{16\eta}{3}\left(\frac{8RT}{\pi M}\right)^{1/2}\frac{P_0}{K_0} \tag{6.89}$$

$$\frac{\varepsilon}{\delta} = \frac{8\eta RTP_0}{{r_p}^2} \tag{6.90}$$

Based on Eqs. (6.86), (6.89), and (6.90), ε/δ, called effective porosity, and r_p, can be obtained as follows.

1. Obtain experimentally permeance P/δ at various average pressure $\overline{p}$.
2. Draw the straight line P/δ versus $\overline{p}$ and obtain the slope P_0 and the intercept with the y axis K_0.
3. Calculate r_p using Eq. (6.89)
4. Calculate ε/δ using Eq. (6.90).

Problem 6.9:

Bakeri et al. (2011) obtained the permeance data of helium gas for his polyetherimide hollow fiber membrane as shown in Table 6.2. Calculate the pore radius and the effective porosity of the membrane.

Answer:

P/δ versus $\overline{p}$ is shown also in Fig. 6.11.

Applying linear regression analysis in Fig. 6.11,

Slope $P_o = 0.7141 \times 10^{-12}$

Intersection $K_0 = 2.7286 \times 10^{-7}$

Using $\eta = 1.98 \times 10^{-5}$ Pa s for helium at 25°C and $M = 4.002 \times 10^{-3}$ kg/mol for helium

$$r_p = \frac{16 \times 1.98 \times 10^{-5}}{3} \times \left(\frac{8 \times 8.314 \times 298.2}{3.1416 \times 4.002 \times 10^{-3}}\right)^{1/2} \times \frac{0.7141 \times 10^{-5}}{2.7286} = 3.471 \times 10^{-7}\text{m}$$

$$\frac{\varepsilon}{\delta} = \frac{8\eta RTP_0}{{r_p}^2} = \frac{8 \times 1.98 \times 10^{-5} \times 8.314 \times 298.2 \times 0.7141 \times 10^{-12}}{(3.471 \times 10^{-7})^2}$$

$$= 2.33 \text{ m}^{-1}$$

Table 6.2 Experimental permeance data obtained by Bakeri et al. (2011).

Feed pressure, bar (gauge)	1	1.5	2	2.5	3	3.5	4
Average pressure $\overline{p}$, bar[a]	1.5	1.75	2	2.25	2.5	2.75	3
Permeance, $\times 10^7$ mol/m^2 s Pa	3.8	3.95	4.2	4.3	4.6	4.6	4.9

[a]Calculated considering the permeate pressure $p_3 = 1$bar, for example, when feed is 1 barg, $\overline{p} = \{(1+1)+1\}/2 = 1.5$bar.

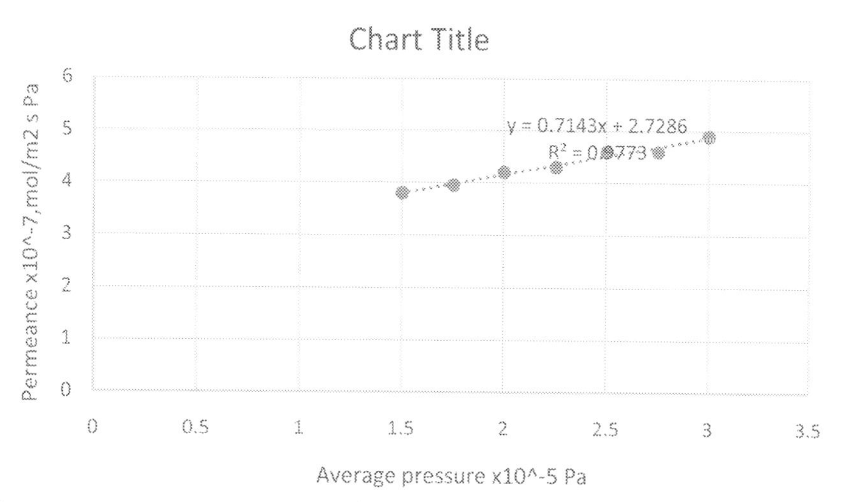

Figure 6.11 Permeance versus average pressure plot.

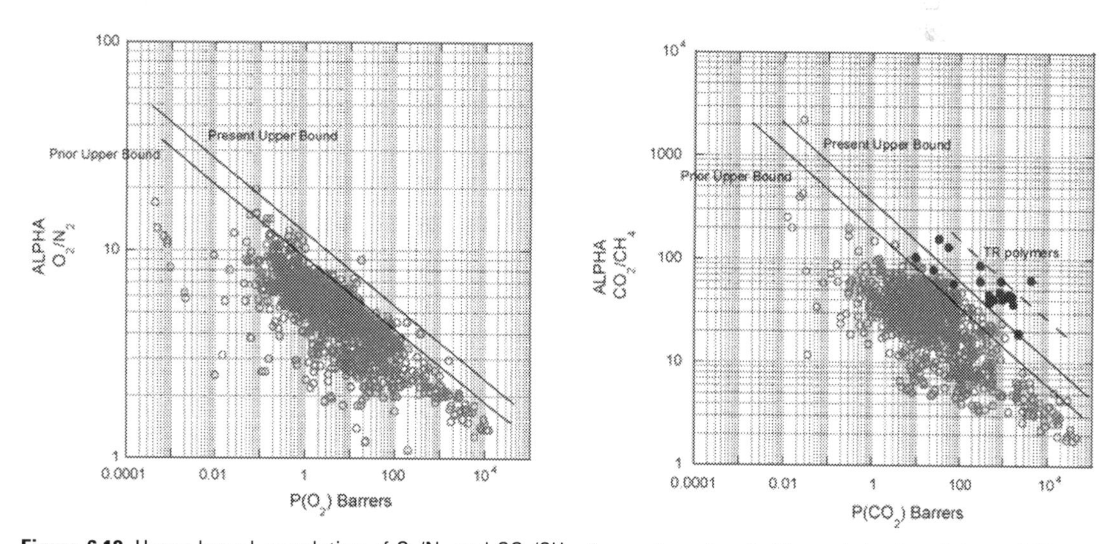

Figure 6.12 Upper bound correlation of O_2/N_2 and CO_2/CH_4. *Source*: Reproduced with permission from Robeson, L.M., 2008. Upper bound revisited. J. Membr. Sci. 320, 390–400.

6.3 Mixed matrix membrane

Despite the success of the application of polymeric membranes for gas separation processes, there is a severe trade-off relationship between permeability and selectivity, as shown by the well known Robeson's plot (Fig. 6.12; Robeson, 2008). After many years of attempts, the boundary line could be shifted only slightly toward the desired right-upward direction. On the other hand, the performance of inorganic membranes has been found to considerably surpass the boundary line of polymeric membranes. Unfortunately, inorganic membranes suffer from poor processability and high cost, which prevent their large-scale applications. Hence, the combination of economical benefit of the processability of the polymeric material and excellent selectivity of the inorganic material was attempted by incorporating inorganic fillers in the host polymeric membrane, which is often called mixed matrix membranes (MMMs). Currently, the most typical filler materials are zeolites, metal oxides, activated carbon, carbon nanotubes, graphene, graphene oxide, clay minerals, metal organic frameworks, etc.

In order to assess the performance improvement of MMMs, many model equations have been derived. Among these, the earliest and most popular is the Maxwell model that is based on the Maxwell equation developed as early as 1873 to evaluate the dielectric properties of the composite material (Maxwell, 1873). The model is written as,

$$P_{eff} = P_c \frac{1 + 2\varphi_d(\alpha - 1)/(\alpha + 2)}{1 - \varphi_d(\alpha - 1)/(\alpha + 2)} \tag{6.91}$$

where P_{eff} and P_c are permeabilities of the MMM and continuous phase (polymer), respectively, φ_d is the volume fraction of the dispersed phase (inorganic fillers), and α is the ratio of the permeability of the dispersed phase (P_d) and the permeability of the continuous phase (P_c).

It should be remembered that the Maxwell model is applicable only for the low filler loading of $\varphi_d < 0.2$. At higher φ_d values, the model cannot predict. As φ_d approaches φ_m, the maximum filler packing density, the deviation becomes more pronounced, especially when $\alpha \rightarrow \infty$.

It should also be remembered that the equation holds only for the ideal case, where polymer tightly adheres to the filler surface without forming any gap between the filler and polymer, while the filler does not change the morphology of the surrounding polymer.

In reality, however, such an ideal case is difficult to achieve. Poor interaction between the filler particle and polymer causes the formation of defects at the filler–polymer interface. The most typical ones are the formation of interfacial void, rigidification of polymer chains, and blockage of the particle pores, as depicted in Fig. 6.13 (Aroon et al., 2010).

In a nonideal case, Eq. (6.91) is no longer applicable. Mahajan (2000) and Vu et al. (2003a,b) assumed the presence of a pseudo-dispersed phase in which the filler is encapsulated in

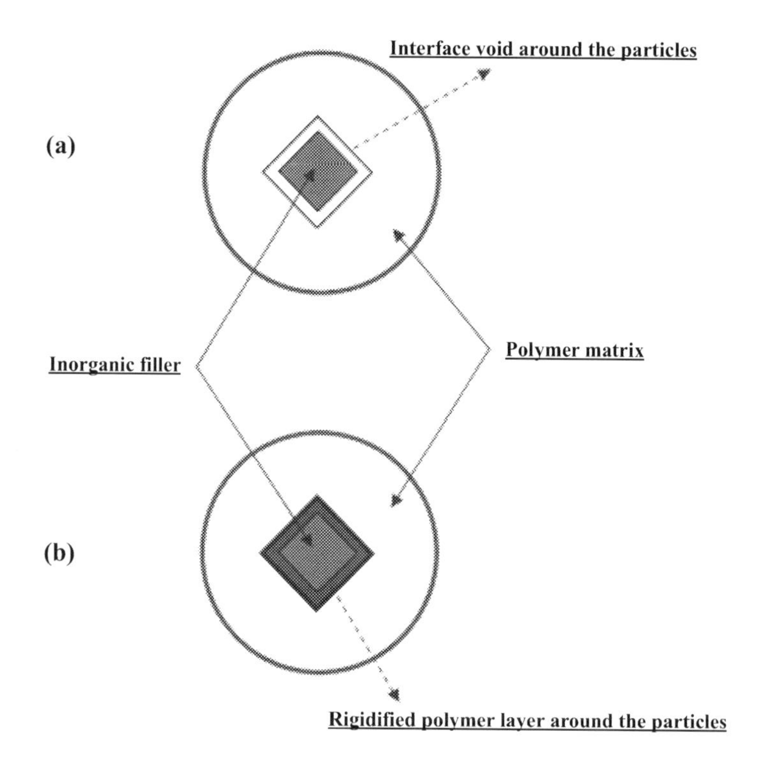

Figure 6.13 Schematic illustration of various MMM morphologies. (a) Interface void and (b) rigidified polymer layer. *Source*: Reproduced with permission from Aroon, M.A., Ismail, A.F., Matsuura, T., Montazer-Rahmati, M.M., 2010. Performance studies of mixed matrix membranes for gas separation: review. Sep. Purif. Technol. 75, 229–242.

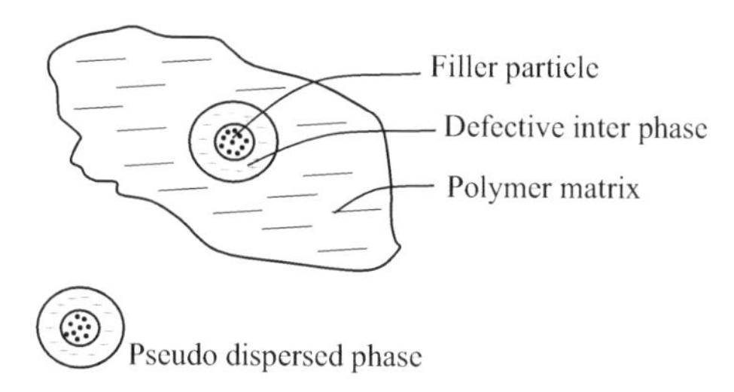

Figure 6.14 Schematic representation of a three-phase model.

an defective interphase. Then the pseudo-dispersed phase is dispersed in the continuous polymer phase. In this way, the three-phase (filler particle, defective interphase, and polymer matrix) system is split into two two-phase systems [pseudo-disperse phase (filler particle + defective interphase) and (pseudo-dispersed phase + polymer matrix)] and the Maxwell equation is applied at each phase Fig. 6.14.

Problem 6.10:

Vu made an MMM membrane which consists of carbon molecular sieve 800-2 (CMS 800-2) and polyimide Matrimid 5218. The following parameters are known.

CO_2 permeability of CMS 800-2, 44.0 Barrer
CO_2 permeability of Matrimid 5218, 10.0 Barrer
CMS 800-2 particle radius, r_{CMS}, 0.5 μm
Thickness of the interphase, t_{int}, 0.075 μm

In the interphase, Matrimid 5218 is rigidified and the permeability becomes 3.33 Barrer.

Calculate the CO_2 permeability of CMS 800-2/Matrimid 5218 MMM when the volume fraction of CMS 800-2 is 33%,

1. For an ideal case
2. For a nonideal case.

Answer:

1. Ideal case
   In Eq. (6.91)

$$\varphi_d = 0.33$$

$$\alpha = \frac{44.0}{10.0} = 4.40$$

Then,

$$P_{eff} = 10.0 \times \frac{1 + 2 \times 0.33 \times (4.40 - 1)/(4.40 + 2)}{1 - 0.33 \times (4.40 - 1)/(4.40 + 2)} = 16.4 \text{ Barrer}$$

2. Nonideal case
   First the permeability of the pseudo-phase is calculated.
   In Eq. (6.91),

$$\varphi_d = \left(\frac{r_{CMS}}{r_{CMS} + t_{int}}\right)^3 = \left(\frac{0.5}{0.5+0.075}\right)^3 = 0.658$$

$$\alpha = \frac{44.0}{3.33} = 13.21$$

Then,

$$P_{eff} = 3.33 \times \frac{1 + 2 \times 0.658 \times (13.21 - 1)/(13.21 + 2)}{1 - 0.658 \times (13.21 - 1)/(13.21 + 2)} = 14.51 \text{ Barrer}$$

Next, the permeability of MMM is calculated
In Eq. (6.91)

$$\varphi_d = \frac{0.33}{0.658} = 0.502$$

$$\alpha = \frac{14.51}{10.0} = 1.451$$

Then,

$$P_{eff} = 10.0 \times \frac{1 + 2 \times 0.502 \times (1.451 - 1)/(1.451 + 2)}{1 - 0.502 \times (1.451 - 1)/(1.451 + 2)} = 12.1 \text{ Barrer}$$

Experimental MMM permeability was 11.5 Barrer. Therefore the nonideal case model is much better than the ideal case model.

Nomenclature

Symbol Definition [dimension (SI unit)]

Symbol	Definition
A	Membrane area (m^2)
c	Concentration (mol/m^3)
d_p	Pore diameter (m)
D	Diffusivity (m^2/s)
D_0	Pre-exponential factor (m^2/s)
E	Potential difference (V)
E_D	Activation energy of diffusion (J/mol)
E_p	Activation energy of permeation (J/mol)
I	Current (A)
J	Flux (mol/m^2 s)
J_k	Flux by Knudsen flow (mol/m^2 s)
J_v	Flux by viscous flow (mol/m^2 s)
k	Boltzmann constant (J/K)
Kn	Knudsen number
M	Molecular weight (kg/mol)
n	Number of pores in unit area of membrane (1 m^{-2})
p	Pressure (Pa)
$\overline{p}$	Average pressure (Pa)
P	Permeability (mol m/m^2 s Pa)
P_{eff}	Permeability of MMM (mol m/m^2 s Pa)
P_c	Permeability of continuous phase (mol m/m^2 s Pa)
P/δ	Permeance (mol/m^2 s Pa)
Q	Permeation rate (mol/s)
r	Pore radius (m)
r_{CMS}	CMS 800-2 particle radius (m)
r_p	Pore radius (m)
R	Gas constant (J/mol K)
R	Resistance in Eq. (6.40) (Ω)
S	Solubility coefficient [(mol/m^3)/Pa]
S_0	Pre-exponential factor [(mol/m^3)/Pa]
t	Time (s)
t_{int}	Thickness of the interphase (m)
T	Temperature (K)
T_g	Glass transition temperature (K)
V	Volume of permeate chamber (m^3)
x	Distance from feed side (m)
x	$= (R_2)_{H_2}$ in Subsection 6.1.4
x with subscript a or b	Mole fraction in feed
X	Mole fraction

y	$= (R_3)_{H_2}$ in Subsection 6.1.4
y with subscript a or b	Mole fraction in permeate
z	$= (R_1')_{H_2}$ in Subsection 6.1.4
Greek letters	
α	Separation factor
α	Ratio of resistances in Subsection 6.1.4
α	P_d/P_c (−)
δ	Membrane thickness (m)
ΔH_s	Enthalpy of solution (J/mol)
ε	Porosity
θ	Time lag (s)
η	Viscosity (Pa s)
λ	Mean free path (m)
σ	Collision diameter (m)
τ	Tortuosity factor
φ	Volume fraction of dispersed phase
Subscripts	
1, 2, 3	Resistance 1, 2, 3 in Subsection 6.1.4
2	Feed
3	Permeate
a	Gas *a*
b	Gas *b*
i	Gas species *i*
sub	Substrate

References

Aroon, M.A., Ismail, A.F., Matsuura, T., Montazer-Rahmati, M.M., 2010. Performance studies of mixed matrix membranes for gas separation: review. Sep. Purif. Technol. 75, 229–242.

Bakeri, Gh, Matsuura, T., Ismail, A.F., 2011. The effect of phase inversion promoters on the structure and performance of polyetherimide hollow fiber membrane using in gas–liquid contacting process. J. Membr. Sci. 383, 159–169.

Crank, J., Park, G.S., 1968. Diffusion in Polymers. Academic Press, London and New York.

Flaconnèche, B., Martin, J., Klopffer, M.H., 2001. Transport properties of gases in polymers: experimental methods. Oil Gas. Sci. Technol. 56, 245–259.

Henis, J.M.S., Tripodi, M.K., 1977. Multicomponent membranes for gas separations. U.S. patent 4230463A.

Khayet, M., Matsuura, T., 2011. Membrane Distillation, Principles and Application. Elsevier, Amsterdam.

Kruczek, B., 1999. Development and characterization of dense membranes for gas separation made from high molecular weight sulfonated poly (phenylene oxide). Effect of casting conditions on the morphology and performance of the membranes. PhD thesis, University of Ottawa.

Li, K., 2007. Ceramic Membranes for Separation and Reaction. John Wiley and Sons, Chichester, UK, p. 73.

Mahajan, R., 2000. Formation, characterization and modeling of mixed matrix membranes materials. PhD dissertation. The University of Texas at Austin.

Matsuura, T., 1994. Synthetic Membranes and Membrane Separation Processes. CRC Press, Florida, Chap 5.
Maxwell, J.C., 1873. Treatise on Electricity and Magnetism. Oxford University Press, London.
Present, R.D., 1958. Kinetic Theory of Gases. McGraw-Hill, New York.
Robeson, L.M., 2008. Upper bound revisited. J. Membr. Sci. 320, 390–400.
Tan, X., Liu, S., Li, K., 2001. Preparation and characterization of inorganic hollow fibre membranes. J. Membr. Sci. 188, 87–95.
Vieth, W.R., 1991. Diffusion in and Through Polymers. Carl Hanser, Munich, Chapter 4).
Vu, D.Q., Koros, W.J., Miller, S.J., 2003a. Mixed matrix membrane using carbon molecular sieves I. Preparation and experimental results. J. Membr. Sci. 211, 311–334.
Vu, D.Q., Koros, W.J., Miller, S.J., 2003b. Mixed matrix membrane using carbon molecular sieves, II, modeling permeation behavior. J. Membr. Sci. 211, 335–348.
Wu, H., 2020. Gas membrane characterization via the time-lag method for neat and mixed-matrix membranes. PhD Thesis, University of Ottawa.

7

Pervaporation

7.1 Pervaporation transport

Pervaporation is a membrane separation process in which the upstream side of the membrane is in contact with feed liquid, while a vacuum is applied on the downstream side of the membrane. The permeant vaporizes somewhere between the upstream and downstream sides of the membrane and the permeate is obtained as vapor. The potential of pervaporation was first mentioned by Binning and coworkers (Binning and James, 1958; Binning et al., 1961). Lee attempted to compare the transport of pervaporation with those of reverse osmosis and gas separation based on the solution-diffusion model under isothermal conditions (Lee, 1975). In his approach, the gradient of the chemical potential across the membrane is considered as the driving force for mass transfer. The chemical potential gradient for species A and B is written as

$$\nabla\mu_A = RT\nabla \ln a_{Am} + \nu_A \nabla p \tag{7.1}$$

$$\nabla\mu_B = RT\nabla \ln a_{Bm} + \nu_B \nabla p \tag{7.2}$$

It is assumed that the pressure, p, remains the same as that of the feed solution throughout the membrane cross-section and falls abruptly at the downstream side from the feed pressure to the permeate pressure, as illustrated in Fig. 7.1 (Matsuura, 1994).

Therefore the second term in Eqs. (7.1) and (7.2) is ignored, which leads to the following flux equations (Lee, 1975).

$$J_A = D_{Am}\frac{c_{Am2} - c_{Am3}}{\delta} \tag{7.3}$$

$$J_B = D_{Bm}\frac{c_{Bm2} - c_{Bm3}}{\delta} \tag{7.4}$$

Thermodynamic equilibrium should be maintained at both sides of the membrane, therefore

$$\mu_{A2} = \mu_{Am2} \tag{7.5}$$

Membrane Separation Processes. DOI: https://doi.org/10.1016/B978-0-12-819626-7.00014-4

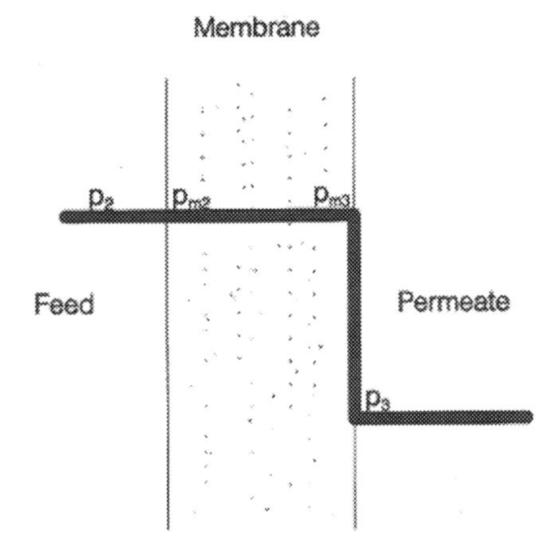

Figure 7.1 Pressure profile in pervaporation (Matsuura, 1994).

$$\mu_{A3} = \mu_{Am3} \tag{7.6}$$

$$\mu_{B2} = \mu_{Bm2} \tag{7.7}$$

$$\mu_{B3} = \mu_{Bm3} \tag{7.8}$$

The subscripts 2 and 3 in the above equations represent the upstream side of the membrane, facing the feed liquid, and the downstream side of the membrane facing the permeate vapor, respectively.

Since

$$\mu = \mu_0 + RT \ln a + \int_{p_{ref}}^{p} \nu dp \tag{7.9}$$

where μ_0 is the chemical potential of pure permeant at $p = p_{ref}$.

Thermodynamic equilibrium at both sides of the membrane for species A and B can be written as

$$a_{Am2} = a_{A2} \exp\left[\frac{-\nu_A(p_{m2} - p_2)}{RT}\right] \tag{7.10}$$

$$a_{Am3} = a_{A3} \exp\left[\frac{-\nu_A(p_{m3} - p_3)}{RT}\right] \tag{7.11}$$

$$a_{Bm2} = a_{B2} \exp\left[\frac{-\nu_B(p_{m2} - p_2)}{RT}\right] \tag{7.12}$$

$$a_{Bm3} = a_{B3} \exp\left[\frac{-\nu_B(p_{m3} - p_3)}{RT}\right] \tag{7.13}$$

with the assumption illustrated in Fig. 7.1.

$$a_{Am2} = a_{A2} \tag{7.14}$$

$$a_{Am3} = a_{A3} \exp\left[\frac{-\nu_A(p_2 - p_3)}{RT}\right] \tag{7.15}$$

$$a_{Bm2} = a_{B2} \tag{7.16}$$

$$a_{Bm3} = a_{B3} \exp\left[\frac{-\nu_B(p_2 - p_3)}{RT}\right] \tag{7.17}$$

Activity is equal to the product of activity coefficient and concentration, hence

$$c_{Am2} = \frac{\gamma_{A2}}{\gamma_{Am2}} c_{A2} \tag{7.18}$$

$$c_{Am3} = \frac{\gamma_{A3}}{\gamma_{Am3}} c_{A3} \exp\left[\frac{-\nu_A(p_2 - p_3)}{RT}\right] \tag{7.19}$$

$$c_{Bm2} = \frac{\gamma_{B2}}{\gamma_{Bm2}} c_{B2} \tag{7.20}$$

$$c_{Bm3} = \frac{\gamma_{B3}}{\gamma_{Bm3}} c_{B3} \exp\left[\frac{-\nu_B(p_2 - p_3)}{RT}\right] \tag{7.21}$$

Defining the partition coefficient, K, as the ratio of the activity coefficients, the above equations become

$$c_{Am2} = K_{A2} c_{A2} \tag{7.22}$$

$$c_{Am3} = K_{A3} c_{A3} \exp\left[\frac{-\nu_A(p_2 - p_3)}{RT}\right] \tag{7.23}$$

$$c_{Bm2} = K_{B2} c_{B2} \tag{7.24}$$

$$c_{Bm3} = K_{B3} c_{B3} \exp\left[\frac{-\nu_B(p_2 - p_3)}{RT}\right] \tag{7.25}$$

Insertion of Eqs. (7.23)–(7.25) into Eqs. (7.3) and (7.4), then yields,

$$J_A = \frac{D_{Am}}{\delta}\left\{K_{A2}c_{A2} - K_{A3}c_{A3}\exp\left[\frac{-\nu_A(p_2 - p_3)}{RT}\right]\right\} \quad (7.26)$$

$$J_B = \frac{D_{Bm}}{\delta}\left\{K_{B2}c_{B2} - K_{B3}c_{B3}\exp\left[\frac{-\nu_B(p_2 - p_3)}{RT}\right]\right\} \quad (7.27)$$

Further,

$$J_A = \frac{P_A}{\delta}\left\{c_{A2} - \alpha_A c_{A3}\exp\left[\frac{-\nu_A(p_2 - p_3)}{RT}\right]\right\} \quad (7.28)$$

$$J_B = \frac{P_B}{\delta}\left\{c_{B2} - \alpha_B c_{B3}\exp\left[\frac{-\nu_B(p_2 - p_3)}{RT}\right]\right\} \quad (7.29)$$

where

$$P_A = D_{AM}K_{A2} \quad (7.30)$$

$$P_B = D_{BM}K_{B2} \quad (7.31)$$

and

$$\alpha_A = \frac{K_{A3}}{K_{A2}} \quad (7.32)$$

$$\alpha_B = \frac{K_{B3}}{K_{B2}} \quad (7.33)$$

Assuming the activity coefficient in the membrane is constant,

$$\alpha_A = \frac{\gamma_{A3}}{\gamma_{A2}} \quad (7.34)$$

$$\alpha_B = \frac{\gamma_{B3}}{\gamma_{B2}} \quad (7.35)$$

Rearranging Eqs. (7.28) and (7.29),

$$J_A = \frac{P_A c_{A2}}{\delta}\left\{1 - \frac{\gamma_{A3}c_{A3}}{\gamma_{A2}c_{A2}}\exp\left[\frac{-\nu_A(p_2 - p_3)}{RT}\right]\right\} \quad (7.36)$$

$$J_B = \frac{P_B c_{B2}}{\delta}\left\{1 - \frac{\gamma_{B3}c_{B3}}{\gamma_{B2}c_{B2}}\exp\left[\frac{-\nu_B(p_2 - p_3)}{RT}\right]\right\} \quad (7.37)$$

while

$$\gamma_A c_A = \frac{p_A}{p_{A^*}} \tag{7.38}$$

$$\gamma_B c_B = \frac{p_B}{p_{B^*}} \tag{7.39}$$

where p_A, p_B, p_{A^*} and p_{B^*} are partial vapor pressure of components A and B, and saturation vapor pressure of components A and B, respectively, and since Eqs. (7.38) and (7.39) should be valid at both sides of the membrane, Eqs. (7.36) and (7.37) can be written as

$$J_A = \frac{P_A c_{A2}}{\delta}\left\{1 - \frac{p_{A3}}{p_{A2}}\exp\left[\frac{-\nu_A(p_2 - p_3)}{RT}\right]\right\} \tag{7.40}$$

$$J_B = \frac{P_B c_{B2}}{\delta}\left\{1 - \frac{p_{B3}}{p_{B2}}\exp\left[\frac{-\nu_B(p_2 - p_3)}{RT}\right]\right\} \tag{7.41}$$

A note regarding the partial vapor pressure is in order. In Eq. (7.40), p_{A3} is the real partial vapor pressure of component A on the permeate side of the membrane since the permeant is in the vapor phase. On the other hand, p_{A2} is the partial vapor pressure of A which is in equilibrium with the feed solution in the liquid phase; that is, p_{A2} the imaginary partial vapor pressure that satisfies Eq. (7.38). The molar volume ν_A is close to that of a liquid, assuming that the permeant is in the liquid phase in the membrane. Then, $-\nu_B(p_2 - p_3)/RT$ in Eqs. (7.40) and (7.41) becomes very small due to the small molar volume of liquid and the exponential term becomes almost equal to unity.

Hence, Eqs. (7.40) and (7.41) can be approximated by

$$J_A = \frac{P_A c_{A2}}{\delta}\left\{1 - \frac{p_{A3}}{p_{A2}}\right\} \tag{7.42}$$

and

$$J_B = \frac{P_B c_{B2}}{\delta}\left\{1 - \frac{p_{B3}}{p_{B2}}\right\} \tag{7.43}$$

When the separation factor is defined

$$\beta = \frac{X_{A3}/X_{A2}}{X_{B3}/X_{B2}} \tag{7.44}$$

And since

$$X_{A3} = \frac{J_A}{J_A + J_B} \tag{7.45}$$

$$X_{B3} = \frac{J_B}{J_A + J_B} \tag{7.46}$$

Eq. (7.44) can be written as

$$\beta = \frac{J_A/c_{A2}}{J_B/c_{B2}} \tag{7.47}$$

Inserting Eqs. (7.42) and (7.43)

$$\beta = \frac{P_A\left(1 - (p_{A3}/p_{A2})\right)}{P_B\left(1 - (p_{B3}/p_{B2})\right)} \tag{7.48}$$

Furthermore, when the pressure on the permeate side is nearly equal to zero,

$$J_A = \frac{P_A c_{A2}}{\delta} \tag{7.49}$$

$$J_B = \frac{P_B c_{B2}}{\delta} \tag{7.50}$$

then,

$$\beta = \frac{P_A}{P_B} \tag{7.51}$$

Problem 7.1:

Often permeability P is based on the transmembrane pressure difference (see Eq. 6.5), instead of the concentration difference. The unit is Barrer, and the pressure based permeability is P'.

Eqs. (7.42) and (7.43) are written as

$$J_A = \frac{P'_A p_{A2}}{\delta}\left\{1 - \frac{p_{A3}}{p_{A2}}\right\} \tag{7.52}$$

$$J_B = \frac{P'_B p_{B2}}{\delta}\left\{1 - \frac{p_{B3}}{p_{B2}}\right\} \tag{7.53}$$

Also, the activity coefficient, γ', is defined based on the mole fraction, X,

$$\gamma_A' X_A = \frac{p_A}{p_{A^*}} \tag{7.54}$$

$$\gamma_B' X_B = \frac{p_B}{p_{B^*}} \tag{7.55}$$

Baker et al. (2010) obtained the following data for the pervaporation of an ethanol/water mixture. Calculate the permeability in Barrer for both ethanol and water.

Permeation rate of ethanol, 4.2 kg/m^2 h;
Permeation rate of water, 8.0 kg/m^2 h;
Separation factor, 5.0.
The operational conditions are:
Feed ethanol concentration, 9.2 wt.%;
Feed temperature, 75°C;
Permeate pressure 5 Torr (=0.5 cmHg).

Answer:

Designate ethanol and water as A and B, respectively.

In 100 g of feed solution ethanol is 9.2 g (9.2/46.07 = 0.200 mol), and water is 90.8 g (90.8/18.02 = 5.039 mol).

Hence,

$$X_{A2} \text{ is } \frac{0.200}{(0.200 + 5.039)} = 0.0382$$

$$X_{B2} \text{ is } \frac{5.039}{(0.200 + 5.039)} = 0.9618$$

Activity coefficients of the ethanol/water mixture are given using an UNIFAC calculator as

$$\text{At } X_{A2} = 0, \ \gamma_{A2} = 7.0009 \text{ and } \gamma_{B2} = 1.0000$$

$$\text{At } X_{A2} = 0.05, \ \gamma_{A2} = 4.6464 \text{ and } \gamma_{B2} = 1.0102$$

By interpolation,

$$\text{At } X_{A2} = 0.0382, \ \gamma_{A2} = 5.202 \text{ and } \gamma_{B2} = 1.008$$

The saturation vapor pressure of ethanol at 75°C (348.15K) is given by the Dortmund data bank as 88.858 kPa (=66.65 cmHg).

The saturation vapor pressure of water at 75°C (348.15K) is given by ENDMEMO as 28.85 cmHg.

From Eqs. (7.54) and (7.55)

$$p_{A2} = (5.202)(0.0382)(66.65) = 13.24 \text{ cmHg}$$

$$p_{B2} = (1.008)(0.9618)(28.85) = 27.97 \text{ cmHg}$$

From Eq. (7.44)

$$\beta = \frac{X_{A3}/X_{B3}}{X_{A2}/X_{B2}} = \frac{X_{A3}/X_{B3}}{0.0382/0.9618} = 5$$

Since $X_{A3} + X_{B3} = 1$

$$X_{A3} = 0.2733 \text{ and } X_{B3} = 0.7267$$

The partial vapor pressures in the permeate are therefore

$$p_{A3} = 0.5 \times 0.2733 = 0.1367 \text{ cmHg}$$

$$p_{B3} = 0.5 \times 0.7267 = 0.3634 \text{ cmHg}$$

Applying Eqs. (7.52) and (7.53),

$$\begin{aligned} P'_A = &[4.2 \text{ kg/m}^2 \text{ h} \times (10^3 \text{ g/1 kg}) \times (1 \text{ mol}/46.02 \text{ g}) \\ &\times (22,400 \text{ cm}^3(\text{STP})/\text{mol})/\{(10^4 \text{ cm}^2/1 \text{ m}^2) \\ &\times (3600 \text{ s/1 h})\} \times 2.5 \, \mu\text{m} \times (10^{-4} \text{ cm}/\mu\text{m})]/(13.24 - 0.1368)\text{cmHg} \\ = &1.083 \times 10^{-6} \text{ cm}^3(\text{STP}) \text{ cm/cm}^2 \text{ s cmHg} = 1.083 \times 10^4 \text{ Barrer} \end{aligned}$$

$$\begin{aligned} P'_B = &[8.0 \text{ kg/m}^2 \text{ h} \times (10^3 \text{ g/1 kg}) \times (1 \text{ mol}/18.02 \text{ g}) \\ &\times (22,400 \text{ cm}^3(\text{STP})/\text{mol})/\{(10^4 \text{ cm}^2/1 \text{ m}^2) \\ &\times (3600 \text{ s/1 h})\} \times 2.5 \, \mu\text{m} \times (10^{-4} \text{ cm}/\mu\text{m})]/(28.03 - 0.42) \text{ cmHg} \\ = &2.500 \times 10^{-6} \text{ cm}^3(\text{STP}) \text{ cm/cm}^2 \text{ s cmHg} = 2.500 \times 10^4 \text{ Barrer} \end{aligned}$$

The ratio of the permeabilities, SF_B^A, is defined as the selectivity. Therefore

$$SF_B^A = \frac{P_A}{P_B} = \frac{1.083 \times 10^4}{2.500 \times 10^4} = 0.433$$

It is interesting to know that the experimental separation factor is more than unity, which means that ethanol is concentrated in the permeate, although the selectivity is less than unity, meaning that the membrane is intrinsically more water selective. This indicates that the increased ethanol concentration in the permeate is largely due to the higher volatility (higher activity coefficient and saturation vapor pressure) of ethanol.

7.2 Pervaporation transport model by Greenlaw and coworkers

Greenlaw et al. discussed pervaporation of a single solvent (Greenlaw et al., 1977). In Lee's derivation, Eqs. (7.3) and (7.4) were obtained by integrating Eqs. (7.1) and (7.2), with the assumption that the pressure and diffusivity are constant across the membrane. It is however known that the diffusivity is highly dependent on the concentration of the permeant in the membrane. Hence, Eqs. (7.3) and (7.4) are written in a differential form, dropping the subscript since only a single solvent is considered.

$$J dx = -D_m dc_m \tag{7.56}$$

where x is the distance in the permeant flow direction from the upstream side of the membrane. [Note that Eqs. (7.3) and (7.4) are obtained by integrating Eq. (7.56) from $x=0$ to $x=\delta$.]

The diffusivity, D_m (m^2/s), is given by Rogers et al. (1960) as

$$D_m = D_{m0}(1 + \alpha c_m^n) \tag{7.57}$$

where D_{m0}, α, and n are constants. The equilibrium sorption is given by the same author as

$$c_m = \sigma\left(\frac{p}{p_*}\right) + \tau\left(\frac{p}{p_*}\right)^{\mathrm{m}} \tag{7.58}$$

where σ, τ, and m are constants and p and p_* are the vapor pressure (Pa) of the permeant (Pa) and the saturation vapor pressure (Pa), respectively. Since the ratio, p/p_*, is equal to the activity of the vapor, Eq. (7.58) may be written as

$$c_m = \sigma a + \tau a^m \tag{7.59}$$

Substituting Eq. (7.57) for D_m of Eq. (7.56) and integrating, we obtain

$$\int_0^\delta J dx = -\int_{c_{2m}}^{c_{3m}} D_{m0}\left(1 + \alpha c_m^n\right) dc_m \tag{7.60}$$

Integration of the above equation yields

$$J\delta = D_{m0}(c_{2m} - c_{3m}) + D_{m0}\frac{\alpha}{n+1}\left(c_{2m}^{n+1} - c_{3m}^{n+1}\right) \tag{7.61}$$

The concentrations c_{2m} and c_{3m} are given as the function of upstream and downstream pressures as follows:

1. For the pure liquid stream at pressure p_2, assuming constant molar volume v, the chemical potential becomes

$$\mu_2 = \mu^* + v(p_2 - p_*) \tag{7.62}$$

where μ^* is the chemical potential of the pure liquid at the saturation vapor pressure.

2. For the permeant dissolved in the membrane at the upstream face,

$$\mu_2 = \mu^* + v(p_2 - p_*) + RT\ln a_2 \tag{7.63}$$

assuming that the molar volume of the permeant in the membrane is the same as that of liquid.

3. For the dissolved permeant at the downstream face,

$$\mu_3 = \mu^* + v(p_2 - p_*) + RT \ln a_3 \tag{7.64}$$

since the assumption was made that the pressure remains constant and is p_2 across the membrane.

4. For pure liquid in contact with the downstream face,

$$\mu_3 = \mu^* + v(p_3 - p_*) \tag{7.65}$$

5. For pure vapor in contact with the downstream face, assuming ideal gas behavior,

$$\mu_3 = \mu^* + RT \ln\left(\frac{p_3}{p_*}\right) \tag{7.66}$$

From Eqs. (7.62) and (7.63)

$$a_2 = 1 \tag{7.67}$$

From Eqs. (7.64) and (7.65)

$$a_3 = \exp\left[-\frac{v}{RT}(p_2 - p_3)\right] \tag{7.68}$$

when $p_3 > p^*$ and the permeate in liquid phase.

From Eqs. (7.64) and (7.66)

$$a_3 = \frac{p_3}{p_*}\exp\left[-\frac{v}{RT}(p_2 - p_*)\right] \tag{7.69}$$

when $p_3 < p_*$ and the permeate is vapor.

Using Eqs. (7.59), (7.61), (7.67) and (7.68) or (7.69), we can calculate permeate flux J.

Problem 7.2:

For the permeation of hexane through a polyethylene film the following parameters are known.

$$D_{m0} = 24.2 \times 10^{-13}\ \text{m}^2/\text{s}$$

$$\alpha = 0.001788$$

$$n = 1.33$$

$$m = 3.4$$

$$\tau = 545.3\ \text{mol/m}^3$$

$$\sigma = 137.9\ \text{mol/m}^3$$

$$p_* = 24{,}918\ \text{Pa}$$

$$v = 130.58 \times 10^{-6}\ \text{m}^3/\text{mol}$$

Calculate the permeation flux of hexane when the film thickness is 2.54×10^{-5} m. The upstream pressure is 101,325 Pa (atmospheric pressure), and the downstream pressure is 40,000 Pa.

Answer:

From Eq. (7.67), $a_2 = 1$; therefore from Eq. (7.59)

$$c_{2m} = 137.9 \times 1 + 545.3 \times 1 = 683.2\ \text{mol/m}^3$$

From Eq. (7.68),

$$a_3 = \exp\left[-\frac{130.58 \times 10^{-6}}{2.479 \times 10^3}\left(1.01325 \times 10^5 - 0.4 \times 10^5\right)\right] = 0.9968$$

$$c_{3m} = 137.9 \times 0.9968 + 545.3 \times 0.9968^{3.4} = 676.9$$

Then, from Eq. (7.61)

$$J = \frac{\left\{\left(24.2 \times 10^{-13}\right)(683.2 - 676.9) + \frac{\left(24.2 \times 10^{-13}\right)(0.001788)}{2.33}\left(683.2^{2.33} - 676.9^{2.33}\right)\right\}}{2.54 \times 10^{-5}}$$

$$= 6.322 \times 10^{-6}\ \text{mol/m}^2\ \text{s}$$

Problem 7.3:

Calculate the flux (mol/m^2 s) when the downstream pressure is 0 and 13,332 Pa.

Answer:

At the downstream pressure of 0 Pa, $a_3 = 0$ and $c_3 = 0$; therefore

$$J = \frac{\left\{(24.2 \times 10^{-13})(683.2) + \frac{(24.2 \times 10^{-13})(0.001788)}{2.33}(683.2^{2.33})\right\}}{2.54 \times 10^{-5}}$$

$$= 35.92 \times 10^{-5}$$

At the downstream pressure of 13,332 Pa, using Eq. (7.69)

$$a_3 = \frac{13,332}{24,918}\exp\left(-\frac{130.58 \times 10^{-6}}{2.479 \times 10^3}(101,325 - 24,918)\right) = 0.5329$$

$$c_{3m} = 137.9 \times 0.5329 + 545.3 \times 0.5329^{3.4} = 137.6$$

Therefore

$$J = \frac{\left\{(24.2 \times 10^{-13})(683.2 - 137.6) + \frac{(24.2 \times 10^{-13})(0.001788)}{2.33}(683.2^{2.33} - 137.6^{2.33})\right\}}{2.54 \times 10^{-5}}$$

$$= 33.91 \times 10^{-5}$$

Problem 7.4:

Calculate the pervaporation flux, mol/m^2 s, when the upstream pressure is 1,206,000 Pa and the downstream pressure is 40,000, 0, and 13,332 Pa.

Answer:

At the downstream pressure of 40,000 Pa,

$$a_3 = \exp\left[-\frac{130.58 \times 10^{-6}}{2.479 \times 10^3}(1,205,768 - 40,000)\right] = 0.9404$$

$$c_{3m} = 137.9 \times 0.9404 + 545.3 \times 0.9404^{3.4} = 572.2$$

Therefore

$$J = \frac{\left\{(24.2 \times 10^{-13})(683.2 - 572.2) + \frac{(24.2 \times 10^{-13})(0.001788)}{2.33}(683.2^{2.33} - 572.2^{2.33})\right\}}{2.54 \times 10^{-5}}$$

$$= 11.01 \times 10^{-5}$$

At the downstream pressure of 0 Pa,

$$J = 35.92 \times 10^{-5}$$

At the downstream pressure of 13,332 Pa,

$$a_3 = \frac{13,332}{24,918}\exp\left(-\frac{130.58 \times 10^{-6}}{2.479 \times 10^3}(1,205,800 - 24,918)\right) = 0.5028$$

$$c_{3m} = 137.9 \times 0.5028 + 545.3 \times 0.5028^{3.4} = 122.0$$

Therefore

$$J = \frac{\left\{(24.2\times10^{-13})(683.2-122.0)+\frac{(24.2\times10^{-13})(0.001788)}{2.33}(683.2^{2.33}-122.0^{2.33})\right\}}{2.54\times10^{-5}}$$

$$= 34.23\times10^{-5}$$

The results of the calculation are summarized in Fig. 7.2.

7.3 A new model for pervaporation transport

A new transport model was developed for the pervaporation of a single component. According to the model, the chemical potential of the permeant is equal to the feed liquid, which is a real liquid phase, at the upstream face of the membrane. Similarly, the chemical potential of the permeant is equal to the permeate vapor, which is a real vapor phase, at the downstream face of the membrane. In between, the presence of an imaginary phase, which turns from liquid to vapor somewhere in the middle of the membrane as the pressure decreases from the upstream to the downstream side of the membrane, is assumed. Note that the assumption of a constant pressure across the membrane is removed in this approach (Fig. 7.3).

Then, at the upstream of the membrane the permeate is in equilibrium with the feed liquid that is under the upstream

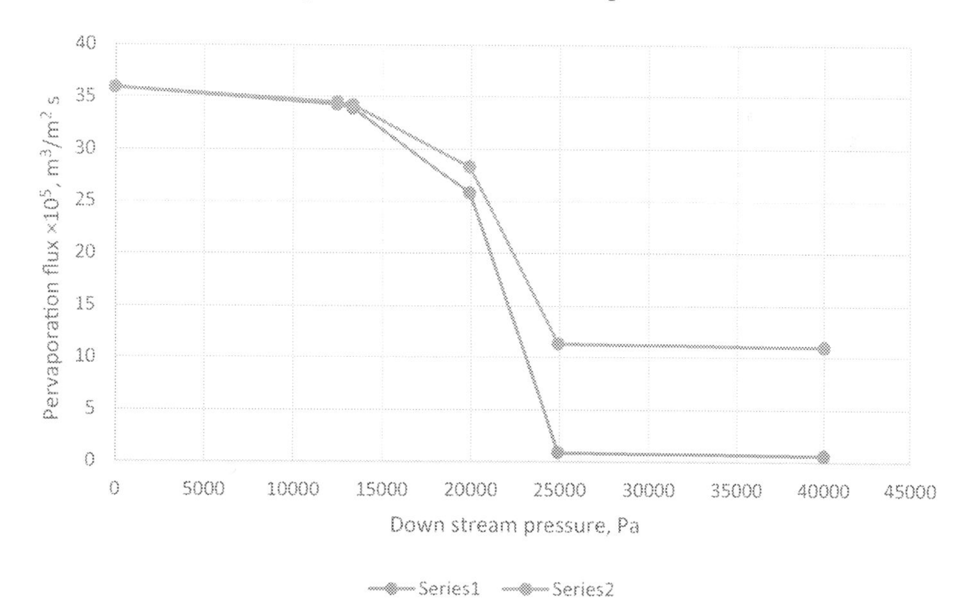

Figure 7.2 Pervaporation flux versus downstream pressure by Greenlaw et al.'s model (series 1, for upstream pressure of 101,325 Pa; series 2, for upstream pressure of 1,206,000 Pa).

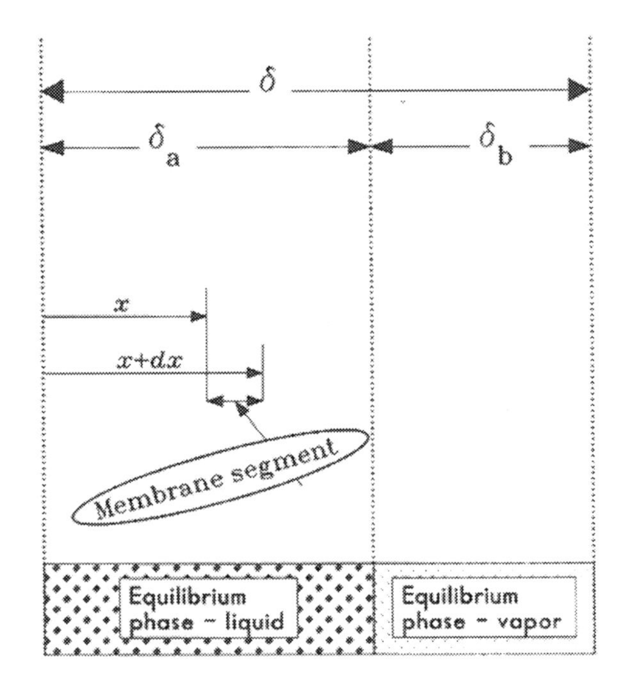

Figure 7.3 Imaginary liquid or vapor phase in equilibrium with permeant in the membrane (Matsuura, 1994).

pressure p_2. The chemical potential of the feed liquid can be written as

$$\mu = \mu^* + \nu(p_2 - p_*) \tag{7.70}$$

where μ^* is the chemical potential of liquid at the saturation vapor pressure p_*. As you move toward the downstream face, with an increase in x in Fig. 7.2, the pressure of the imaginary liquid phase decreases. The change of the chemical potential, $d\mu$, in a distance segment dx is then given by

$$d\mu = \nu dp \tag{7.71}$$

where the molar volume of liquid, ν, is a constant for incompressible liquid and dp is the pressure change in the imaginary liquid phase. Eventually, the pressure in the imaginary phase becomes the saturation pressure p_* at $x = \delta_a$ (see Fig. 7.2), and thereafter the imaginary phase becomes vapor.

Applying

$$J = -\frac{c_m}{f_m}\frac{d\mu}{dx} = -\frac{c_m}{f_m}\frac{\nu dp}{dx} \tag{7.72}$$

where c_m and f_m are the permeant concentration in the membrane and the friction to the movement of the permeant in the membrane, respectively. [The explanation of Eq. (7.72) is in order. $\frac{c_m}{f_m}$ in the equation is equal to the phenomenological coefficient of the irreversible thermodynamics. Lonsdale used this equation to derive Eq. (3.3) of Chapter 3: Reverse Osmosis, Forward Osmosis, and Pressure-Retarded Osmosis, with an assumption that $\frac{c_m}{f_m}$ is constant.]

Integrating Eq. (7.72), from $x = 0$ to $x = \delta_a$, where imaginary phase is liquid,

$$J \int_0^{\delta_a} dx = -\frac{c_m}{f_m} \nu \int_{p_2}^{p_*} dp \quad (7.73)$$

and

$$J = -\frac{c_m}{f_m} \frac{\nu(p_* - p_2)}{\delta_a} \quad (7.74)$$

Further using the relationship,

$$f_m = \frac{RT}{D_m} \quad (7.75)$$

where D_m is the diffusivity in the membrane.

The flux, J, becomes,

$$J = \frac{D_m c_m}{RT} \frac{\nu(p_2 - p_*)}{\delta_a} \quad (7.76)$$

Since the vapor pressure of the imaginary liquid phase is the same as the saturation vapor pressure p_*, c_m, which is in equilibrium with the imaginary phase, is according to Eq. (7.58)

$$c_m = \sigma + \tau \quad (7.77)$$

Further from Eq. (7.57)

$$D_m = D_{m0}\{1 + \alpha(\sigma + \tau)^n\} \quad (7.78)$$

Insertion of Eqs. (7.77) and (7.78) into Eq. (7.76) yields

$$J = \frac{\left(\frac{D_{m0}}{RT}\right)(\sigma + \tau)\{1 + \alpha(\sigma + \tau)^n\}\nu}{\delta_a}(p_2 - p_*) \quad (7.79)$$

From δ_a to δ the imaginary phase is vapor.

Applying the ideal gas law, $\nu = \frac{RT}{p}$, the change in chemical potential in the imaginary phase can be written as

$$d\mu = \frac{RT}{p} dp \quad (7.80)$$

Then,

$$J = -\frac{c_m}{f_m}\frac{d\mu}{dx} = -\frac{c_m}{f_m}\frac{RT}{p}\frac{dp}{dx} \quad (7.81)$$

Rearranging,

$$Jdx = -\frac{c_m RT}{f_m}\frac{dp}{p} \tag{7.82}$$

Using Eqs. (7.57) and (7.58)

$$Jdx = -\frac{\{\sigma(p/p_*) + \tau(p/p_*)^m\}RT}{\left[RT/D_{m0}\left\{1+\alpha\left(\sigma(p/p_*)+\tau(p/p_*)^m\right)^n\right\}\right]}\frac{d(p/p_*)}{(p/p_*)} \tag{7.83}$$

Rearranging,

$$Jdx = -\frac{D_{m0}\{\sigma(p/p_*) + \tau(p/p_*)^m\}\left\{1+\alpha\left(\sigma(p/p_*)+\tau(p/p_*)^m\right)^n\right\}}{(p/p_*)}d(p/p_*) \tag{7.84}$$

Setting $\zeta = p/p_*$,

$$Jdx = -\frac{D_{m0}\{\sigma\zeta + \tau\zeta^m\}\{1+\alpha(\sigma\zeta+\tau\zeta^m)^n\}}{\zeta}d\zeta \tag{7.85}$$

Integrating from $x = \delta_a$ to δ,

$$J\int_{\delta_a}^{\delta} dx = -\int_{p_*/p_*}^{p_3/p_*}\frac{D_{m0}\{\sigma\zeta + \tau\zeta^m\}\{1+\alpha(\sigma\zeta+\tau\zeta^m)^n\}}{\zeta}d\zeta \tag{7.86}$$

And

$$J = \frac{\int_{p_3/p_*}^{1}\frac{D_{m0}\{\sigma\zeta + \tau\zeta^m\}\{1+\alpha(\sigma\zeta+\tau\zeta^m)^n\}}{\zeta}d\zeta}{\delta_b} \tag{7.87}$$

where

$$\delta_b = \delta - \delta_a \tag{7.88}$$

Furthermore, from Eq. (7.79)

$$\delta_a = \frac{\left(\frac{D_{m0}}{RT}\right)(\sigma+\tau)\{1+\alpha(\sigma+\tau)^n\}\nu}{J}(p_2 - p_*) \tag{7.89}$$

From Eq. (7.87)

$$\delta_b = \frac{\int_{p_3/p_*}^{1}\frac{D_{m0}\{\sigma\zeta + \tau\zeta^m\}\{1+\alpha(\sigma\zeta+\tau\zeta^m)^n\}}{\zeta}d\zeta}{J} \tag{7.90}$$

Since $\delta_a + \delta_b = \delta$

$$\delta = \frac{\left(\frac{D_{m0}}{RT}\right)(\sigma+\tau)\{1+\alpha(\sigma+\tau)^n\}\nu}{J}(p_2 - p_*) + \frac{\int_{p_3/p_*}^{1}\frac{D_{m0}\{\sigma\zeta + \tau\zeta^m\}\{1+\alpha(\sigma\zeta+\tau\zeta^m)^n\}}{\zeta}d\zeta}{J} \tag{7.91}$$

Therefore

$$J = \frac{\left(\frac{D_{m0}}{RT}\right)(\sigma + \tau)\{1 + \alpha(\sigma+\tau)^n\}\nu}{\delta}(p_2 - p_*)$$

$$+ \frac{\int_{p_3/p_*}^{1} \frac{D_{m0}\{\sigma\zeta + \tau\zeta^m\}\{1+\alpha(\sigma\zeta+\tau\zeta^m)^n\}}{\zeta} d\zeta}{\delta} \quad (7.92)$$

Eq. (7.92) is valid when $p_3 < p_*$. When $p_3 > p_*$
The flux will be

$$J = \frac{\left(\frac{D_{m0}}{RT}\right)(\sigma + \tau)\{1 + \alpha(\sigma+\tau)^n\}\nu}{\delta}(p_2 - p_3) \quad (7.93)$$

Problem 7.5:

Use the same numerical values as those given in Problem 7.2 and calculate the pervaporation flux when the upstream pressure is 101,325 Pa and the downstream pressure is 0, 12,466, 19,934, and 40,000 Pa.

Answer:
For the downstream pressure of 0 Pa, Eq. (7.92) is used, since $0 < 24{,}918$. The flux, J, is as follows.

$$J = \frac{\left(\frac{24.2\times10^{-13}}{2.479\times10^{3}}\right)\times(137.9+545.3)\times\{1+0.001788\times(137.9+545.3)^{1.33}\}\times(130.58\times10^{-6})\times(101{,}325-24{,}918)}{2.54\times10^{-5}}$$

$$+\left(\frac{24.2\times10^{-13}}{2.54\times10^{-5}}\right)\int_0^{\zeta}\frac{(137.9\zeta+545.3\zeta^{3.4})\times\left\{1+0.001788\times(137.9\zeta+545.3\zeta^{3.4})^{1.33}\right\}}{\zeta}d\zeta$$

$$= 13.843\times10^{-5}\ \text{mol/m}^2\ \text{s}$$

Similarly, for the downstream pressures of 12,466 and 19,934 Pa, the pervaporation fluxes are $12{,}753\times10^{-5}$ and 9.295×10^{-5} mol/m^2 s, respectively.

When the downstream pressure is 40,000 Pa, $40{,}000 > 24{,}918$. Therefore from Eq. (7.93)

$$J = \frac{\left(\frac{24.2\times10^{-13}}{2.479\times10^{3}}\right)\times(137.9+545.3)\times\{1+0.001788\times(137.9+545.3)^{1.33}\}\times(130.58\times10^{-6})\times(101{,}325-40{,}000)}{2.54\times10^{-5}}$$

$$= 0.243\times10^{-5}\ \text{mol/m}^2\text{s}$$

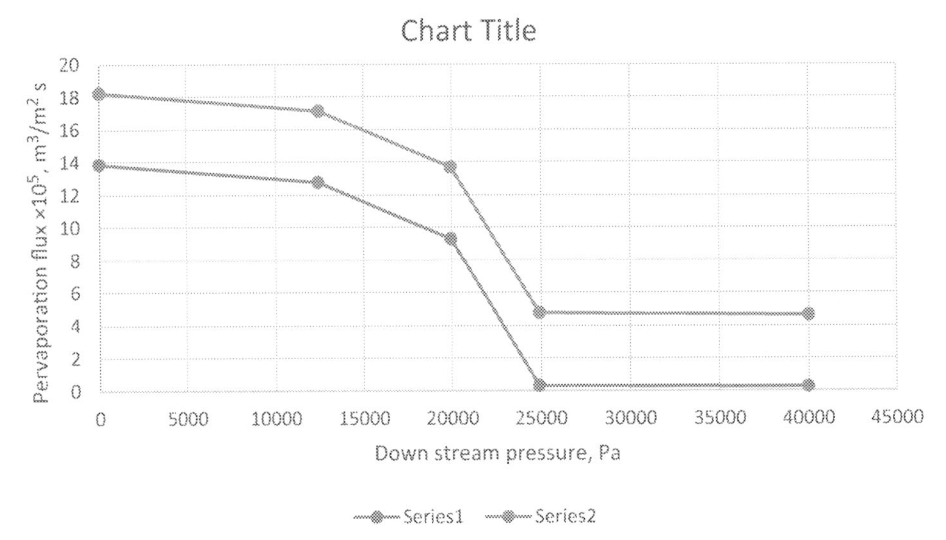

Figure 7.4 Pervaporation flux versus downstream pressure by the new model (series 1, for upstream pressure of 101,325 Pa; series 2, for upstream pressure of 1,206,000 Pa).

Problem 7.6:

Calculate pervaporation fluxes when the upstream pressure is 1.206×10^6 Pa for the downstream pressures of 0, 12,466, 19,934, and 40,000 Pa.

The answers are as follows:

Downstream pressure, Pa	Pervaporation flux $\times 10^5$, mol/m^2 s
0	18.226
12,466	17.136
19,934	13.678
40,000	4.626

The results of the calculation by the new model are summarized in Fig. 7.4.

Note that the pervaporation flux at zero downstream pressure increases with an increase in the upstream pressure according to the new model, whereas by using the model of Greenlaw et al., the upstream pressure has no effect on the pervaporation flux. Also, fluxes calculated by the new model are lower than those calculated by Greenlaw et al.'s model.

Nomenclature

Symbol

$\boldsymbol{a}$	Activity (mol/m^3)
c	Concentration (mol/m^3)
$\boldsymbol{D}$	Diffusivity (m^2/s)
D_{m0}	Constant defined in Eq. (7.57) (m^2/s)
f_m	Friction against the movement of permeant [(J/mol)/(m^2/s)]
$\boldsymbol{J}$	Flux (mol/m^2 s)
$\boldsymbol{K}$	Partition coefficient
$\boldsymbol{m}$	Constant defined in Eq. (7.58)
$\boldsymbol{n}$	Constant defined in Eq. (7.57)
$\boldsymbol{p}$	Pressure (Pa)
$\boldsymbol{p}_{ref}$	Reference pressure (Pa)
p_*	Saturation vapor pressure (Pa)
$\boldsymbol{P}$	Permeability (mol m/m^2 s Pa)
$\boldsymbol{R}$	Gas constant (J/mol K)
SF_B^A	Selectivity
$\boldsymbol{T}$	Absolute temperature (K)
x	Distance from the upstream side (m)
$\boldsymbol{X}$	Mole fraction
Greek letters	
α	Ratio of partition coefficient
α	Constant defined in Eq. (7.57) [(mol/m^3)$^{-n}$]
β	Separation factor
γ	Activity coefficient
δ	Membrane thickness (m)
δ_a	Length of imaginary liquid phase (m)
δ_b	Length of imaginary vapor phase (m)
ζ	$= p/p_*$
μ	Chemical potential (J/mol)
μ_0	Chemical potential of pure permeant (J/mol)
μ^*	Chemical potential of pure liquid at the saturation vapor pressure (J/mol)
ν	Molar volume (m^3/mol)
σ	Constant defined in Eq. (7.58) (mol/m^3)
τ	Constant defined in Eq. (7.58) (mol/m^3)
Subscripts	
2	Upstream side
3	Downstream side
$\boldsymbol{A}$	Species A
$\boldsymbol{B}$	Species B
$\boldsymbol{m}$	In the membrane

References

Baker, R.W., Wijmans, J.G., Huang, Y., 2010. Permeability, permeance and selectivity: a preferred way of reporting pervaporation performance data. J. Membr. Sci. 348, 346–352.

Binning, R.C., James, F.E., 1958. Now separate by membrane permeation. Petrol. Refin. 37, 214.

Binning, R.C., Lee, R.J., Jennings, J.F., Martin, E.C., 1961. Separation of liquid mixtures by permeation. Ind. Eng. Chem. 53, 45–50.
Greenlaw, F.W., Prince, W.D., Shelden, R.A., Thompson, E.V., 1977. Dependence of diffusive permeation rates on upstream and downstream pressures: single component permeant. J. Membr. Sci. 2, 141–151.
Lee, C.H., 1975. Theory of reverse osmosis and some other membrane permeation operations. J. Appl. Polym. Sci. 19, 83–95.
Matsuura, T., 1994. Synthetic Membranes and Membrane Separation Processes. CRC Press, Boca Raton, FL (Chapter 5).
Rogers, C.E., Stannett, V., Szwarz, M., 1960. The sorption, diffusion and permeation of organic vapors in polyethylene. J. Polym. Sci. 45, 61–82.

8

Membrane distillation

8.1 About membrane distillation

8.1.1 Process principles

Membrane distillation (MD) is a process mainly suited for applications in which water is the major component of the feed solution. MD is a thermally driven process in which only vapor molecules are transported through porous hydrophobic membranes from the hot feed side to the cold permeate side. The liquid feed to be treated by MD is maintained in direct contact with one side of the membrane (feed side) without penetrating into the dry pores. The hydrophobic property of the membrane prevents liquid feed from entering its pores due to the surface tension forces. Thus, the transport mechanism of MD consists of:

- Evaporation of water at the warm feed side of the membrane;
- Migration of water vapor through the nonwetted pores;
- Condensation of water vapor transported at the permeate side (Fig. 8.1).

Simultaneous heat and mass transfer occur in MD and various applications (desalination, environmental/waste cleanup, water reuse, food, medical, etc.) by different MD configurations, such as direct contact MD (DCMD), sweeping gas MD, vacuum MD, and air gap membrane MD (AGMD).

Since MD is a thermally driven process, it requires the supply of heat. However, unlike the conventional distillation process, heat of lower quality (i.e., lower temperature) can be used in MD, which enables MD to be operated with waste heat and/or alternative energy sources such as solar and geothermal energy. Furthermore, the requirement for lower hydrostatic pressures than in pressure-driven processes such as reverse osmosis (RO), nanofiltration (NF), ultrafiltration (UF), and microfiltration (MF), the less demanding mechanical properties of the membrane, and high solute rejection achievable especially during the treatment of water that contains nonvolatile solutes make MD more attractive than any other membrane separation processes (Qtaishat, 2008).

Membrane Separation Processes. DOI: https://doi.org/10.1016/B978-0-12-819626-7.00013-2

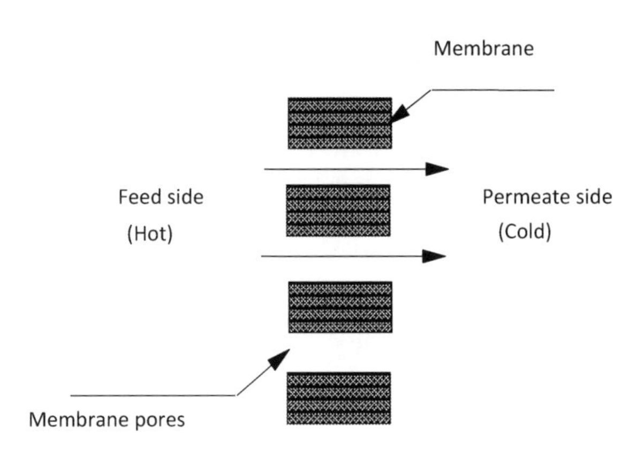

Figure 8.1 Principle of MD (Qtaishat, 2008). *MD*, Membrane distillation.

8.1.2 Different membrane distillation configurations

There are typically four different MD configurations, depending on the method by which vapor is removed once it has migrated through the pores (Qtaishat, 2008).

8.1.2.1 Direct contact membrane distillation

In DCMD, liquid is in direct contact with the membrane at both the feed and permeate sides. The feed temperature is maintained higher than that of the permeate. Vapor migrates through the dry pore from the feed to the permeate side due to the vapor pressure difference.

The vapor diffusion path is limited by the thickness of the membrane, thereby reducing mass and heat transfer resistances. Condensation within the pores is avoided by selecting appropriate temperature differences across the membrane.

8.1.2.2 Air gap membrane distillation

In AGMD, there is a stagnant air gap between the permeate side of the membrane and the condenser surface where the vapor condenses to liquid. Vapor should travel through the membrane pores and air gap before it reaches the condenser, which increases heat and mass transfer resistances. Although heat loss by conduction is reduced, flux is reduced also. Larger temperature differences can be applied across the membrane, which can compensate in part for the greater transfer resistances.

8.1.2.3 Vacuum membrane distillation

Vapor on the feed side is transported through the pores by applying a vacuum on the permeate side. The permeate side

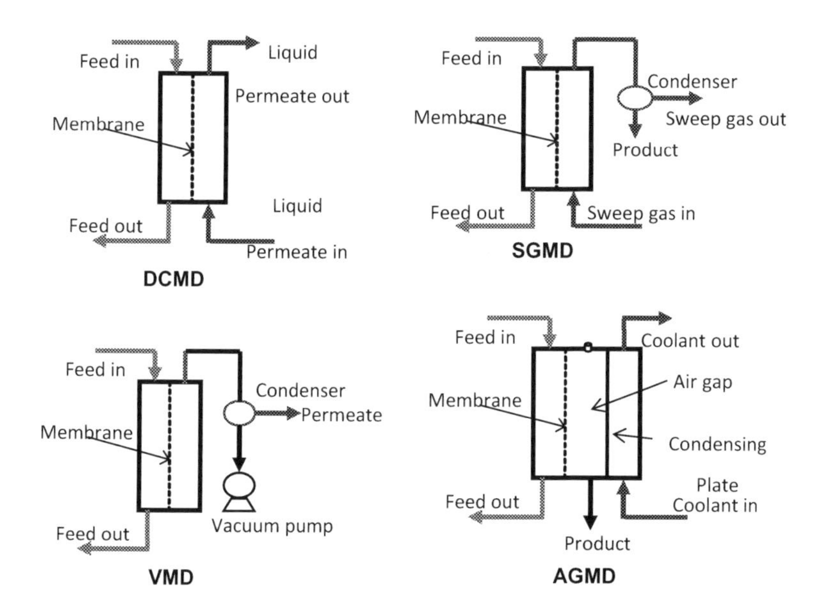

Figure 8.2 Four configurations of membrane distillation (Qtaishat, 2008).

pressure is maintained at lower than the saturation vapor pressure of the permeant and the condensation of the vapor takes place outside the membrane module.

8.1.2.4 Sweep gas membrane distillation

The permeating vapor is removed using an inert gas stream which passes on the permeate side of the membrane. The vapor is condensed outside the membrane module. This configuration involves large volumes of the sweep gas and vapor stream.

Fig. 8.2 shows the different configurations of MD.

8.2 Transport in direct contact membrane distillation

Herein the MD transport theory is described for DCMD in detail. The transport equations for other MD configurations can be derived with some modifications (Qtaishat, 2008).

8.2.1 Heat transfer

In DCMD, the heat transfer occurs in three steps, as described in Fig. 8.3:

1. Heat transfer through the feed boundary layer;
2. Heat transfer through the membrane;

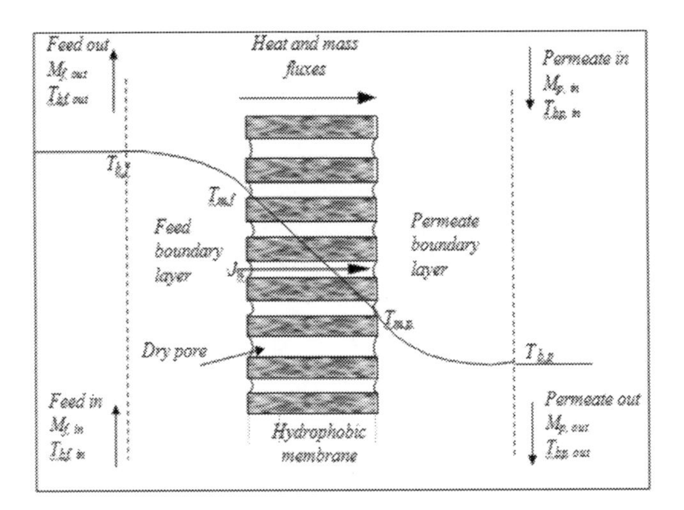

Figure 8.3 Heat and mass transfer in MD (Qtaishat, 2008). *MD*, Membrane distillation.

3. Heat transfer through the permeate boundary layer.

The heat transfer through the feed boundary, Q_f (J/m^2 s), is given by

$$Q_f = h_f(T_{b,f} - T_{m,f}) \tag{8.1}$$

where $T_{b,f}$ (K) and $T_{m,f}$ (K) are the temperature of bulk liquid and at the membrane surface on the feed side, respectively (see Fig. 8.3), and h_f (J/m^2 s K) is the heat transfer coefficient of the feed boundary layer.

The heat transfer through the membrane, Q_m (J/m^2 s), is given as the sum of the heat conduction, Q_c (J/m^2 s), through the membrane and the latent heat, Q_v (J/m^2 s), carried by the vapor when it travels through the membrane pore.

Hence,

$$Q_m = Q_C + Q_v \tag{8.2}$$

Q_C is given by

$$Q_C = \frac{k_m}{\delta}(T_{m,f} - T_{m,p}) \tag{8.3}$$

where k_m (J/m s K) is the thermal conductivity of the membrane, δ (m) is the thickness of the membrane, and $T_{m,p}$ (K) is the temperature at the membrane surface on the permeate side (see Fig. 8.3).

Further, k_m is calculated by the following equation.

$$k_m = \varepsilon k_g + (1 - \varepsilon)k_p \tag{8.4}$$

where ε is the porosity of the membrane, and k_g (J/m s K) and k_p (J/m s K) are the thermal conductivity of the gas filling the membrane pore and the thermal conductivity of the membrane material (usually polymer), respectively.

Q_V in Eq. (8.2) is given by

$$Q_V = J_w \Delta H_{v,w} \tag{8.5}$$

where J_w (kg/m^2 s) is the flux of vapor (in most cases water) and $\Delta H_{v,w}$ (J/kg) is the heat of vaporization of the liquid (in most cases water).

The heat transfer through the permeate boundary layer, Q_p (J/m^2 s), is given by

$$Q_p = h_p(T_{m,p} - T_{b,p}) \tag{8.6}$$

$T_{b,p}$ (K) and $T_{m,p}$ (K) are the temperatures of bulk liquid and at the membrane surface on the permeate side, respectively (see Fig. 8.3), and h_p (J/m^2 s K) is the heat transfer coefficient of the permeate boundary layer.

Since

$$Q_f = Q_m = Q_p \tag{8.7}$$

$$h_f(T_{b,f} - T_{m,f}) = \frac{k_m}{\delta}(T_{m,f} - T_{m,p}) + J_w \Delta H_{v,w} = h_p(T_{m,p} - T_{b,p}) \tag{8.8}$$

Setting

$$Q = Q_f = Q_m = Q_p \tag{8.9}$$

and written as

$$Q = H(T_{b,f} - T_{b,p}) = H\Delta T \tag{8.10}$$

where H is the overall heat transfer coefficient,

$$Q = \left[\frac{1}{h_f} + \frac{1}{\frac{k_m}{\delta} + \frac{J_w \Delta H_{v,w}}{(T_{m,f} - T_{m,p})}} + \frac{1}{h_p}\right]^{-1} (T_{b,f} - T_{b,p}) \tag{8.11}$$

and

$$H = \left[\frac{1}{h_f} + \frac{1}{\frac{k_m}{\delta} + \frac{J_w \Delta H_{v,w}}{(T_{m,f} - T_{m,p})}} + \frac{1}{h_p}\right]^{-1} \tag{8.12}$$

The temperature polarization coefficient defined as

$$\theta = \frac{T_{m,f} - T_{m,p}}{T_{b,f} - T_{b,p}} \tag{8.13}$$

is the ratio of the temperature difference across the membrane, which actually causes the vapor pressure difference across the membrane, to the temperature difference between the bulk feed and permeate stream. In the ideal case, θ is equal to unity but in reality it is less than unity.

When θ is less than 0.2, DCMD is heat transfer limited due to poor module design. When θ is higher than 0.6, DCMD is mass transfer limited with low membrane permeability.

8.2.2 Mass transfer

The mass transfer in DCMD is in most cases the transfer of water vapor, and the mass flux is proportional to the water vapor pressure difference between two ends (feed and permeate side) of the pore, that is, it is given by (Qtaishat, 2008),

$$J_w = B_m(p_{m,f} - p_{m,p}) \tag{8.14}$$

where J_w is water flux, B_m is permeance, $p_{m,f}$ and $p_{m,p}$ are the partial pressures of water at the feed and permeate sides evaluated by using Antoine equation at the temperatures $T_{m,f}$ and $T_{m,p}$, respectively, such that

$$p^v = \exp\left(23.328 - \frac{3841}{T - 45}\right) \tag{8.15}$$

where p^v is the water vapor pressure in Pa and T is the corresponding temperature in K (Qtaishat, 2004).

Various types of mechanisms have been proposed for transport of gases or vapors through porous membranes, such as the Knudsen model, viscous model, ordinary-diffusion model, and sometimes these flow resumes are combined. Which mechanism is operative under a given experimental condition depends on the Knudsen number, K_n, defined as the ratio of the mean free path (λ) of the transported molecules to the pore size (diameter, d_p) of the membrane; that is, $K_n = \lambda/d_p$. Since in the DCMD process both the hot feed and the cold permeate water are brought into contact with the membrane under atmospheric pressure, the total pressure is constant at ≈ 1 atm and viscous flow becomes negligible (Phattaranawik et al., 2003; Khayet et al., 2004).

Therefore, mass transport across the membrane occurs in DCMD in three regions depending on the pore size and mean free path of the transferring species (Phattaranawik et al., 2003): Knudsen region, continuum region (or ordinary-diffusion region), and transition region (or combined Knudsen/ordinary-diffusion region).

If the mean free path of transporting water molecules is large compared to the membrane pore size (i.e., $K_n > 1$ or $r < 0.5\lambda$, where r is pore radius), the molecule–pore wall collisions are dominant over the molecule–molecule collisions (see Fig. 6.10) and the mass transfer occurs in the Knudsen region.

In this case the permeance is given by (Qtaishat, 2008)

$$B_m^K = \frac{2\varepsilon r}{3\tau\delta}\left(\frac{8M}{\pi RT}\right)^{1/2} \tag{8.16}$$

where ε, τ, r, and δ are the porosity, pore tortuosity factor, pore radius, and thickness of the hydrophobic membrane, respectively; M is the molecular weight of water, R is the gas constant, and T is the absolute temperature.

If the mean free path of transporting water molecules is smaller than the pore size (i.e., $K_n < 0.01$ or $r > 50\lambda$), molecular diffusion is used to describe the mass transport in the continuum region caused by the virtually stagnant air trapped within each membrane pore due to the low solubility of air in water. Then, the following relationship can be used for the net DCMD membrane permeability (Khayet et al., 2004).

$$B_M^D = \frac{\varepsilon pD}{\tau\delta p_a RT} \tag{8.17}$$

where p_a is the air pressure, p is the total pressure inside the pore assumed to be constant and equal to the sum of the partial pressures of air and water vapor, and D is the water diffusivity, wherein pD (Pa m^2/s) for water–air is calculated from the following expression (Phattaranawik et al., 2003).

$$pD = 1.895\ 10^{-5}T^{2.072} \tag{8.18}$$

In the transition region ($0.01 < K_n < 1$ or $0.5\lambda < r < 50\lambda$), water molecules collide with each other and diffuse in air. The mass transport occurs via the combined Knudsen/ordinary-diffusion mechanism and the following equation can be used to calculate the water permeance (Phattaranawik et al., 2003; Khayet et al., 2004).

$$B_m^C = \left[\frac{3\tau\delta}{2\varepsilon r}\left(\frac{\pi RT}{8M}\right)^{1/2} + \frac{\tau\delta P_a RT}{\varepsilon PDM}\right]^{-1} \tag{8.19}$$

Problem 8.1:

A DCMD experiment was conducted with feed pure water at the feed and permeate side temperature ($T_{b,f}$ and $T_{b,p}$) of 50°C and 20°C, respectively. A water vapor flux, J_w, of 2.0×10^{-3} kg/m^2 s was obtained. Calculate the feed and permeate temperatures

($T_{m,f}$ and $T_{m,p}$) at the membrane surfaces and the temperature polarization factor, θ, using the following parameters:

Heat of vaporization of water ($\Delta H_{v,w}$) = 2405.55 kJ/kg;

Heat transfer coefficient on the feed side (h_f) = 2000 W/m^2 K;

Heat transfer coefficient on the permeate side (h_p) = 2000 W/m^2 K;

Thermal conductivity of gas-filled membrane (k_m) = 0.02 W/m K;

Membrane thickness (δ) = 50 μm.

Answer:

Using the given data

$$J_w\,\Delta H_{v,w} = \left(2.0\times 10^{-3}\right)\times\left(2405.55\times 10^{3}\right) = 4811.1\ \mathrm{J/m^2\,s}$$

From Eqs. (8.7)–(8.9), it is obvious that

$$Q = \frac{k_m}{\delta}\left(T_{m,f} - T_{m,p}\right) + J_w\Delta H_{v,w} \tag{8.20}$$

Then, from Eq. (8.11)

$$\mathrm{Q} = \left[\frac{1}{h_f} + \frac{T_{m,f} - T_{m,p}}{Q} + \frac{1}{h_p}\right]^{-1}\times\left(T_{b,f} - T_{b,p}\right) \tag{8.21}$$

$$\mathrm{Q}\times\left[\frac{1}{h_f} + \frac{T_{m,f} - T_{m,p}}{Q} + \frac{1}{h_p}\right] = \left(T_{b,f} - T_{b,p}\right)$$

$$\mathrm{Q}\times\left(\frac{1}{h_f} + \frac{1}{h_p}\right) + \left(T_{m,f} - T_{m,p}\right) = \left(T_{b,f} - T_{b,p}\right)$$

Inserting Eq. (8.20)

$$\left(\frac{k_m}{\delta}\left(T_{m,f} - T_{m,p}\right) + J_w\Delta H_{v,w}\right)$$

$$\times\left(\frac{1}{h_f} + \frac{1}{h_p}\right) + \left(T_{m,f} - T_{m,p}\right) = \left(T_{b,f} - T_{b,p}\right)$$

$$\left(T_{m,f} - T_{m,p}\right)\times\left\{\frac{k_m}{\delta}\times\left(\frac{1}{h_f} + \frac{1}{h_p}\right) + 1\right\}$$

$$= \left(T_{b,f} - T_{b,p}\right) - J_w\Delta H_{v,w}\times\left(\frac{1}{h_f} + \frac{1}{h_p}\right)$$

$$\left(T_{m,f} - T_{m,p}\right) = \frac{\left(T_{b,f} - T_{b,p}\right) - J_w\Delta H_{v,w}\times\left(\frac{1}{h_f} + \frac{1}{h_p}\right)}{\frac{k_m}{\delta}\times\left(\frac{1}{h_f} + \frac{1}{h_p}\right) + 1}$$

Inserting the given parameters

$$(T_{m,f} - T_{m,p})$$

$$= \frac{(323.2 - 293.2) - \left((2.0 \times 10^{-3}) \times (2405.55 \times 10^{3})\right) \times \left(\frac{1}{2000} + \frac{1}{2000}\right)}{\frac{0.02}{50 \times 10^{-6}} \times \left(\frac{1}{2000} + \frac{1}{2000}\right) + 1}$$

$$= 18.00 \quad (8.22)$$

From Eq. (8.8)

$$h_f\left(T_{b,f} - T_{m,f}\right) = h_p\left(T_{m,p} - T_{b,p}\right)$$

Inserting the numerical values,

$$2000 \times \left(323.2 - T_{m,f}\right) = 2000 \times \left(T_{m,p} - 293.2\right)$$

Rearranging

$$(T_{m,f} + T_{m,p}) = 614.4 \quad (8.23)$$

Solving Eqs. (8.22) and (8.23) simultaneously,
$T_{m,f}$ and $T_{m,p}$ are 317.2K and 299.2K, respectively.
Then,

$$\theta = \frac{317.2 - 299.2}{323.2 - 293.2} = 0.6$$

Problem 8.2:

DCMD experiments were carried out with feed pure water at the feed and permeate side temperatures ($T_{b,f}$ and $T_{b,p}$) of 50°C and 20°C, respectively, using a membrane with the following specifications:

Membrane porosity (ε) = 0.1921;
Membrane pore radius (r) = 11.43×10^{-9} m;
Membrane thickness (δ) = 50.92×10^{-6} m;
Thermal conductivity of gas-filled membrane (k_m): 0.02 W/m K.

Calculate the flux (J_w), the feed and permeate temperatures ($T_{m,f}$ and $T_{m,p}$) at the membrane surface and the temperature polarization factor, θ.

Heat transfer coefficient on the feed side (h_f) = 2000 W/m^2 K.

Heat transfer coefficient on the permeate side (h_p) = 2000 W/m^2 K.

Answer:

First, the mean free path (λ) is calculated by

$$\lambda = \frac{RT}{\sqrt{2}\pi\sigma^2 N_A p} \quad (8.24)$$

where R is the gas constant (8.314 J/K mol), σ is the collision diameter (2.641×10^{-10} m for water vapor) and N_A is the Avogadro number (6.022×10^{23} 1/mol). T and p are the temperature (K) and pressure (Pa) in the pore, respectively. Using the average temperature [(50 + 20)/2 = 35°C] and the atmospheric pressure (1.01325×10^5 Pa),

$$\lambda = \frac{8.314 \times 308.2}{1.4142 \times 3.1416 \times (2.641 \times 10^{-10})^2 \times (6.022 \times 10^{23}) \times (1.01325 \times 10^5)}$$

$$= 1.355 \times 10^{-7}\text{m}$$

Hence,

$$K_n = \frac{\lambda}{d_p} = \frac{1.355 \times 10^{-7}}{2 \times 11.43 \times 10^{-9}} = 5.93 > 1$$

Therefore, the mass transport occurs in the Knudsen region. Then, according to Eq. (8.16), permeability (B_m^K) is given by,

$$B_m^K = \frac{2 \times 0.1921 \times (11.43 \times 10^{-9})}{3 \times 1 \times (50.92 \times 10^{-6})} \times \left\{ \frac{8 \times (18.02 \times 10^{-3})}{3.1416 \times 8.314 \times 308.2} \right\}^{1/2}$$

$$= 1.217 \times 10^{-7}\text{s/m}$$

From Eqs. (8.14) and (8.15)

$$J_w = B_m \left(\exp\left(23.328 - \frac{3841}{T_{m,f} - 45} \right) - \exp\left(23.328 - \frac{3841}{T_{m,p} - 45} \right) \right)$$

Eq. (8.23) is also applicable for this case, hence

$$T_{m,p} = 614.4 - T_{m,f}$$

And

$$J_w = B_m \left(\exp\left(23.328 - \frac{3841}{T_{m,f} - 45} \right) - \exp\left(23.328 - \frac{3841}{569.4 - T_{mf}} \right) \right) \tag{8.25}$$

Also, from Eqs. (8.19) and (8.20)

$$\frac{k_m}{\delta}(T_{m,f} - T_{m,p}) + J_w \Delta H_{v,w}$$

$$= \left[\frac{1}{h_f} + \frac{T_{m,f} - T_{m,p}}{\frac{k_m}{\delta}(T_{m,f} - T_{m,p}) + J_w \Delta H_{v,w}} + \frac{1}{h_p} \right]^{-1} \times (T_{b,f} - T_{b,p})$$

Again applying Eq. (8.23)

$$\frac{k_m}{\delta}\left(2T_{m,f}-614.4\right)+J_w\Delta H_{v,w}$$

$$=\left[\frac{1}{h_f}+\frac{2T_{m,f}-614.4}{\frac{k_m}{\delta}\left(2T_{m,f}-614.4\right)+J_w\Delta H_{v,w}}+\frac{1}{h_p}\right]^{-1}$$

$$\times\left(T_{b,f}-T_{b,p}\right) \quad (8.26)$$

Inserting all known parameters in Eqs. (8.25) and (8.26)

$$J_w=\left(1.217\times10^{-7}\right)\times\left(\exp\left(23.328-\frac{3841}{T_{m,f}-45}\right)\right.$$

$$\left.-\exp\left(23.328-\frac{3841}{569.4-T_{mf}}\right)\right) \quad (8.27)$$

$$\frac{0.02}{50.92\times10^{-6}}\left(2T_{m,f}-614.4\right)+J_w\times\left(2405.55\times10^3\right)$$

$$=\left[\frac{1}{2000}+\frac{2T_{m,f}-614.4}{\frac{0.02}{50.92\times10^{-6}}\left(2T_{m,f}-614.4\right)+J_w\times\left(2405.55\times10^3\right)}+\frac{1}{2000}\right]^{-1}$$

$$\times(323.2-293.2) \quad (8.28)$$

From Eqs. (8.27) and (8.28), two unknowns, J_w and $T_{m,f}$, can be obtained.

The answers are

$$J_w=8.29\times10^{-4}\ \text{kg/m}^2\ \text{s}$$

and $T_{m,f}$ = 317.3 K and $T_{m,p}$ = 297.1 K.

$$\theta=\frac{317.3-297.1}{323.2-293.2}=0.673$$

Problem 8.3:

The vapor pressure of aqueous sodium chloride is given, instead of Eq. (8.15), as

$$p^v=\exp\left(23.328-\frac{3841}{T-45}\right)\times\gamma_w\times X_w \quad (8.29)$$

where $\gamma_w$andX_w are, respectively, the activity coefficient and mole fraction of water.

Furthermore, γ_w can be obtained by

$$\gamma_w = 1 - 0.5X_{\text{NaCl}} - 10X_{\text{NaCl}}^2 \tag{8.30}$$

where X_{NaCl} is the mole fraction of NaCl.

What will the flux be when the feed is 3.5 wt.% NaCl solution under the same experimental conditions as in Problem 8.2?

Answer:

The mole fraction of NaCl in the feed aqueous solution is $X_{\text{NaCl}} = \frac{(3.5/58.45)}{\left(\frac{3.5}{58.45}\right) + \left(\frac{100-3.5}{18.02}\right)} = 0.0111.$

From Eq. (8.30)

$$\gamma_w = 1 - 0.5 \times 0.0111 - 10 \times 0.0111^2 = 0.9932$$

From Eq. (8.29) the vapor pressure of the feed aqueous solution is

$$p^v = \exp\left(23.328 - \frac{3841}{T-45}\right) \times 0.9932 \times 0.9889$$

$$= 0.9822 \times \exp\left(23.328 - \frac{3841}{T-45}\right)$$

Eq. (8.15) can be used for the permeate side, since the permeate is pure water.

Then, Eq. (8.27) becomes

$$J_w = (1.217 \times 10^{-7}) \times \left(0.9822 \times \exp\left(23.328 - \frac{3841}{T_{m,f} - 45}\right) - \exp\left(23.328 - \frac{3841}{569.4 - T_{mf}}\right)\right) \tag{8.31}$$

And Eqs. (8.27) and (8.31) are solved simultaneously for J_w and $T_{m,f}$.

There is practically no change in $T_{m,f}$ and $J_w = 8.09 \times 10^{-4}$ m^3/m^2 s.

Nomenclature

Symbol	Definition [dimension (SI unit)]
B_m	Permeance (kg/m^2 s Pa)
B_m^K	Permeance of Knudsen diffusion (kg/m^2 s Pa)
B_M^D	Permeance of ordinary diffusion (kg/m^2 s Pa)
B_M^C	Permeance of combined Knudsen and ordinary diffusion (kg/m^2 s Pa)
D	Diffusivity (m^2/s)
h	Heat transfer coefficient (J/m^{2}s K)

H	Overall heat transfer coefficient (J/m^2 s K)
J_w	Flux of water (kg/m^2 s)
k_g	Thermal conductivity of the gas filling the membrane pore (J/m s K)
k_m	Thermal conductivity of membrane (J/m s K)
k_p	Thermal conductivity of membrane material (J/m s K)
M	Molecular weight (kg/mol)
N_A	Avogadro number (1/mol)
p	Vapor pressure (Pa)
p	Total pressure inside the pore in Eq. (8.17) (Pa)
p^v	Saturation vapor pressure (Pa)
Q	Heat flux (J/m^2 s)
Q_c	Heat flux by conduction (J/m^2 s)
Q_m	Heat flux through membrane (J/m^2 s)
Q_v	Heat flux by latent heat (J/m^2 s)
R	Gas constant (J/mol K)
T	Temperature (K)

Greek letters

δ	Membrane thickness (m)
$\Delta H_{v,w}$	Heat of vaporization of water (J/kg)
ΔT	Temperature difference between bulk feed and bulk permeate (K)
X_{NaCl}	Mole fraction of NaCl
X_w	Mole fraction of water

Greek letters

γ_w	Activity coefficient of water
ε	Membrane porosity
θ	Temperature polarization coefficient
σ	Collision diameter (m)
τ	Tortuosity factor

Subscripts

b	Bulk liquid
f	Feed side
m	Membrane surface
p	Permeate side

References

Khayet, M., Velázquez, A., Mengual, J.I., 2004. Modelling mass transport through a porous partition: effect of pore size distribution. J. Non-Equil. Thermodyn. 29, 279–299.

Phattaranawik, J., Jiraratananon, R., Fane, A.G., 2003. Effect of pore size distribution and air flux on mass transport in direct contact membrane distillation. J. Membr. Sci. 215, 75–85.

Qtaishat, M.R., 2004. Use of Vacuum Membrane Distillation for Concentrating Sugars and Dyes from Their Aqueous Solutions (M.Sc. thesis). Jordan University of Science and Technology, Jordan.

Qtaishat, M.R., 2008. Design of Novel Membranes for Desalination by Membrane Distillation (Ph.D. thesis). Department of Chemical Engineering, University of Ottawa.

9

Membrane contactor (membrane absorption) and membrane adsorption

9.1 Membrane contactor

The development of industries worldwide increases the emission of hazardous materials to the environment, which causes climatic changes (Bakeri et al., 2010, 2012). It is believed that the emission of greenhouse gases such as CO_2 increases the global temperature, which can cause disasters such as flooding and drought. On the other hand, the emission of dangerous materials from industries can also have an effect on human health; for example, in petrochemical industries, the production of polycarbonate resin requires use of phosgene which may be harmful at concentration levels as low as below 1 ppm. Therefore emissions of this gas in the plant exit gas should be low, necessitating its removal, for example, by absorption with a caustic (NaOH) solution. Also, in ammonia plants, the emission of ammonia to the atmosphere is reduced by absorption with water. In some cases, the component that should be removed is not a hazardous material but one that can cause some operational problems, for example, water should be separated from natural gas because it may condense or freeze in pipelines and damage them.

In the conventional absorption process, absorption towers are used for such separations in which an absorbent flows countercurrently in a packed or tray tower. However, these towers have some operational disadvantages such as:

1. Low contact area;
2. Weeping, entrainment, and flooding;
3. Dependency of liquid and gas flow rates;
4. High liquid loss because of solvent evaporation.

An alternative to separation towers is a membrane contactor in which gas and liquid flow in opposite sides of a porous membrane; the solute gas diffuses from the bulk of gas to the entrance of a pore and by diffusion through the pore of the membrane to

Membrane Separation Processes. DOI: https://doi.org/10.1016/B978-0-12-819626-7.00010-7

the other end of the pore, and is absorbed by the liquid. A membrane contactor can provide a high and well-defined contact area per unit volume of equipment, and is reported to be 4–30 times more effective than conventional separation columns. Also, a membrane contactor can reduce the capital investment and operating costs of absorption processes, as reported elsewhere.

Compared to gas separation membranes, a membrane contactor has some advantages and disadvantages. In the gas separation membrane, the process is a single step, but in the membrane contactor, the absorbed component(s) should be separated from absorbent in a stripping process; resulting in a two-step process. On the other hand, the transmembrane pressure difference in the gas separation membrane is much higher than that in the membrane contactor, and more complicated design and module fabrication are required for gas separation. Furthermore, the flux per unit area in the membrane contactor is much higher than in the gas separation membrane due to the difference in diffusion mechanism. In the gas separation membrane, the gas should dissolve in the dense skin layer of the membrane and diffuse through it. The gas should further diffuse through the gas-filled pores of the sublayer. On the other hand, in the membrane contactor, the gas should diffuse through the gas-filled pores of the membrane. As the diffusivity in the gas phase is much higher than in the solid phase, the membrane contactor has a much higher flux compared to the gas separation membrane.

The selectivity of the membrane contactor is governed by the type of absorbent and the relative solubility of the gas components in the absorbent. Therefore, the selectivity of the membrane contactor is much higher than in the gas separation membrane, for example, the solubility of CO_2 and CH_4 in water, the simplest absorbent for separation of CO_2, at 25°C and 1 bar are 1.7 g/kg water and 0.023 g/kg water, respectively. In other words, the selectivity of the absorption process is almost 74.

The pores of the membrane in contactor applications should be gas filled, as the diffusivity in the gas phase is 10^4 times higher than diffusivity in the liquid phase. The penetration of liquid into the membrane pores should be prevented as pore wetting reduces the mass transfer in the contactor significantly and makes it less competitive compared to the conventional column. One cause of pore wetting is capillary condensation, but, more importantly, the pressure of the feed liquid should surpass a critical value for the liquid to enter into the liquid pores. This critical value, called the liquid entry pressure of water, depends on some properties of the membrane such as pore size, hydrophobicity, surface roughness, and chemical resistance to solvent and also, on the surface tension of the solvent and operating conditions of the absorption process. Thus, it is possible to reduce the wettability of membranes by decreasing the pore size and using membranes with a highly hydrophobic surface.

9.1.1 Transport in membrane contactor

With their large surface area to volume ratio, hollow fibers are most suitable for the membrane contactor. In the hollow fiber gas flows on the shell side and liquid flows on the lumen side (or vice versa) and the gas (called solute gas, CO_2 for example) diffuses through the gas-filled pores of the membrane from the shell side and is absorbed by liquid at the mouth of the pores on the lumen side. Therefore, the following three resistances are connected in series in the membrane contactor: (1) resistance in the concentration boundary layer on the gas side, (2) membrane resistance, and (3) resistance in the concentration boundary layer on the liquid side. Although membrane creates extra resistance in the membrane contactor, a large contact area that the membrane provides for mass transfer compensates for this extra resistance. The mass transfer path in a membrane contactor is shown in Fig. 9.1.

Among those, the resistance at the gas boundary layer is usually ignored because the gas diffusivity is very high.

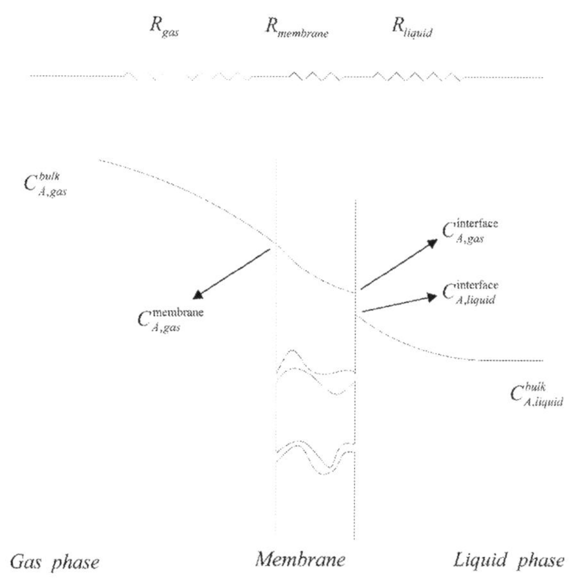

Figure 9.1 Mass transfer resistances in series for the nonwetted gas–liquid membrane contactor (Bakeri et al., 2010).

Regarding the solute gas transport through the membrane pores, it is either by the Knudsen diffusion mechanism, the bulk diffusion mechanism, or a combination of both, depending on the pore size. If the pore diameter (d_p) < 100 nm, Knudsen diffusion is predominant, if $d_p > 10\ \mu m$ bulk diffusion is predominant, and between these two ranges, both mechanisms exist (Kumar et al., 2003; Li and Chen, 2005). Although the diffusion regime is primarily governed by the pore size at the skin layer, the mass transfer rate through a membrane is not dependent solely on the skin layer. The pore diameter in the sublayer of asymmetric membranes is much larger than that in the skin layer and bulk diffusion is predominant in the sublayer. Also, the thickness of the sublayer is much greater than the thickness of the skin layer, and so the sublayer may also have resistance to mass transfer. After diffusing through the pore and when the gas comes into contact with liquid at the pore mouth on the permeate (lumen) side, the absorbent liquid is assumed to be saturated instantly with CO_2 gas. Also, unlike gas phase resistance, the resistance in the liquid phase boundary layer cannot be ignored.

The overall mass transfer coefficient (K_{OL}) is then given by:

$$\frac{1}{K_{OL}} = \frac{1}{k_L} + \frac{Hd_i}{k_m d_{lm}} + \frac{Hd_i}{k_g d_o} \tag{9.1}$$

where k_L is liquid side mass transfer coefficient, H is Henry's constant, k_m is membrane mass transfer coefficient, k_g is gas side mass transfer coefficient, and d_i, d_o, and d_{lm} are inner diameter, outer diameter, and log mean diameter of hollow fiber membrane, respectively. The last term disappears by ignoring the gas side resistance.

The overall mass transfer coefficient (K_{OL}) can be calculated by

$$K_{OL} = \frac{Q_L\left(C_l^{out} - C_l^{in}\right)}{A\Delta C_l^{av}} \tag{9.2}$$

where K_{OL} is overall mass transfer coefficient (m/s), Q_L is liquid flow rate (m^3/s), C_l is solute gas (CO_2) concentration in liquid (mol/m^3), where the superscripts *out* and *in* indicate hollow fiber outlet and inlet, respectively, and ΔC_l^{av} is logarithmic mean of transmembrane concentration difference of solute gas in terms of liquid (mol/m^3), which can be calculated by Eq. (9.3). A is the contact area (m^2) which is calculated based on the inner diameter of hollow fiber membrane as liquid flows in the lumen side (see Fig. 9.2).

$$\Delta C_l^{av} = \frac{\left(HC_g^{in} - C_l^{out}\right) - \left(HC_g^{out} - C_l^{in}\right)}{ln\left(\frac{HC_g^{in} - C_l^{out}}{HC_g^{out} - C_l^{in}}\right)} \tag{9.3}$$

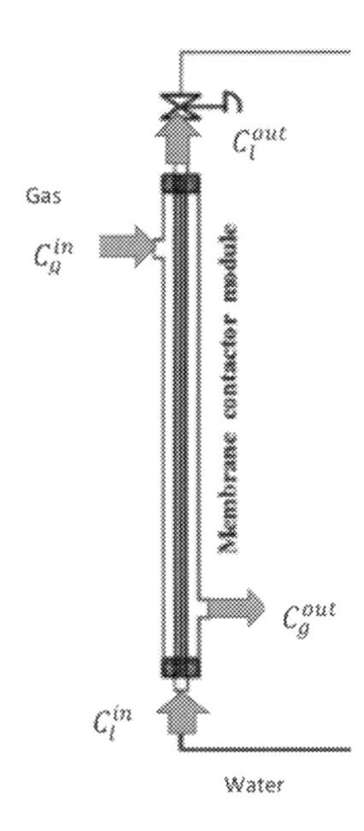

Figure 9.2 Membrane contactor device.

Note that the solute gas concentration in the gas phase C_g is converted to HC_g, the concentration of solute gas in the liquid, which is in equilibrium with the gas. Thus, ΔC_l^{av} expresses the driving force for the solute gas transport.

Dindore et al. (2004), on the other hand, presented Eq. (9.4) to calculate the overall mass transfer coefficient (K_{OL}):

$$J_{av} = \frac{Q_L m C_g \left(1 - exp\left(\frac{-K_{OL}\pi d_i L}{Q_L}\right)\right)}{\pi d_i L} \tag{9.4}$$

where J_{av} is average absorption flux, m is distribution factor which is 0.85 at 25°C, and L is length of hollow fiber membrane (m). Hence, K_{OL} can be calculated either by Eqs. (9.3) or (9.4) using the experimental data.

In Eq. (9.1), the liquid side mass transfer coefficient, k_L, can be obtained from the Sherwood number, Sh, which is correlated to the Reynolds, Re, and Schmidt number, Sc, for a tube by

$$Sh = 1.62\left(Re \times Sc \times \frac{d_i}{L}\right)^{0.33} \text{ for laminar flow} \tag{9.5}$$

$$Sh = 0.04R_e^{0.75} \times Sc^{0.33} \text{ for turbulent flow} \tag{9.6}$$

and

$$Sh = \frac{k_L}{D/d_i} \tag{9.7}$$

$$Re = \frac{d_i v_L \rho}{\mu} \tag{9.8}$$

$$Sc = \frac{\mu}{\rho D} \tag{9.9}$$

where D is the diffusivity of gas in the liquid, v_L is the liquid velocity in the lumen side of the tube, and ρ and μ are the density and viscosity of the liquid, respectively.

Kumar et al. (2003) proposed to calculate Sh using the Graetz number, Gr, in the following equations.

$$Sh = 3.67 \text{ for } Gz < 10 \tag{9.10}$$

$$Sh = 1.62 Gz^{1/3} \text{ for } Gz > 20 \tag{9.11}$$

Or, for the entire range of Gz, the following equations can be used (Kreulen et al., 1993):

$$Sh = \sqrt[3]{3.67^3 + 1.62^3 Gr} \tag{9.12}$$

$$Gz = \frac{v_L d_i^2}{DL} \tag{9.13}$$

9.1.2 Wilson plot

This is a method to evaluate k_L and k_m from a simple plot of $1/K_{OL}$ versus v_L^{-m}.

In Eq. (9.1), ignoring the third term of the right hand side (gas boundary layer resistance),

$$\frac{1}{K_{OL}} = \frac{1}{k_L} + \frac{Hd_i}{k_m d_{im}} \tag{9.14}$$

Combining Eqs. (9.5), (9.7)–(9.9), and (9.14)

$$\frac{1}{K_{OL}} = \frac{1}{1.62 d_i^{-0.34} D^{0.67} L^{-0.33}} v_L^{-0.33} + \frac{Hd_i}{k_m d_{im}} \text{ for laminar flow region} \tag{9.15}$$

And combining Eqs. (9.6)–(9.9), and (9.14)

$$\frac{1}{K_{OL}} = \frac{1}{0.04 d_i^{-0.25} \rho^{0.42} \mu^{-0.42} D^{0.67}} v_L^{-0.75} + \frac{Hd_i}{k_m d_{im}} \text{ for turbulent flow region} \tag{9.16}$$

In both cases,

$$\frac{1}{K_{OL}} = a\, v_L^{-m} + b \tag{9.17}$$

Therefore plotting a straight line, $1/K_{OL}$ versus $v_L{}^{-m}$, the intercept with the y-axis, C, is $Hd_i/k_m d_{im}$

Then, the membrane mass transfer coefficient $k_m = Hd_i/d_{im}C$

Thus, Wilson's method is useful to separate liquid boundary resistance and membrane resistance.

Problem 9.1:

Bakeri et al. fabricated polyetherimide hollow fiber membrane (inner diameter 0.45 mm, length 19 cm) and used it in a membrane contactor for the absorption of feed CO_2 gas by water (Bakeri et al., 2010). CO_2 gas was supplied to the shell side of the membrane at 1 bar gauge, while the water flowed in the lumen side at 1.5 bar gauge, to prevent the formation of CO_2 bubbles in the lumen side, at a flow velocity of 0.54 m/s.

They observed the CO_2 permeation rate of 2.025×10^{-3} mol/m^2 s. Calculate the overall mass transfer coefficient.

Answer:

In Eq. (9.4)

$$J_{av} = \frac{Q_L m C_g \left(1 - exp\left(\frac{-K_{OL}\pi d_i L}{Q_L}\right)\right)}{\pi d_i L}$$

$$d_i = 0.45 \times 10^{-3}\ \text{m}$$

$$L = 0.19\ \text{m}$$

$$Q_L = \frac{0.54 \times 3.1416 \times \left(0.45 \times 10^{-3}\right)^2}{4} = 0.0859 \times 10^{-6}\text{m}^3/\text{s}$$

$$C_g = \frac{2 \times 10^5}{8.314 \times 298.2} = 80.67\ \text{mol/m}^3$$

From Eq. (9.4)

$$K_{OL} = \frac{-Q_L}{\pi d_i L}\ln\left(1 - \frac{J_{av}\pi d_i L}{Q_L m C_g}\right) = \frac{-0.0859\times 10^{-6}}{3.1416\times(0.45\times 10^{-3})\times 0.19}$$

$$\ln\left\{1 - \frac{(2.025\times 10^{-3})\times 3.1416\times(0.45\times 10^{-3})\times(0.19)}{0.0859\times 10^{-6}\times 0.85\times 80.67}\right\}$$

$$= 3.099\times 10^{-5}\,\mathrm{m/s}$$

Problem 9.2:

Bakeri et al. obtained K_{OL} for different v_L values. The data are summarized in Table 9.1 (Bakeri et al., 2010). Obtain the resistance of the membrane by applying the Wilson plot.

Answer:

$$\frac{1}{K_{OL}} \text{ vs } v_L^{-0.75} \text{ plot is tried.}$$

Table 9.2 shows $v_L{}^{-0.75}$ and $1/K_{OL}$. A $1/K_{OL}$ versus $v_L{}^{-0.0.75}$ plot is shown in Fig. 9.3.

The plot shows a good linear relationship with $R^2 = 0.9789$. From the intercept at the y axis, 0.1586×10^5 s/m is obtained as the resistance of the membrane.

Table 9.1 K_{OL} **for different** v_L **values.**

v_L, m/s	$K_{OL}\times 10^5$, m/s
0.14	1.75
0.17	1.85
0.32	2.50
0.54	3.00
0.76	3.95

Table 9.2 $1/K_{OL}$ **versus** $v_L{}^{-0.75}$.

$v_L{}^{-0.75}$	$1/K_{OL}\times 10^{-5}$
4.369	0.571
3.777	0.541
2.350	0.400
1.587	0.333
1.228	0.255

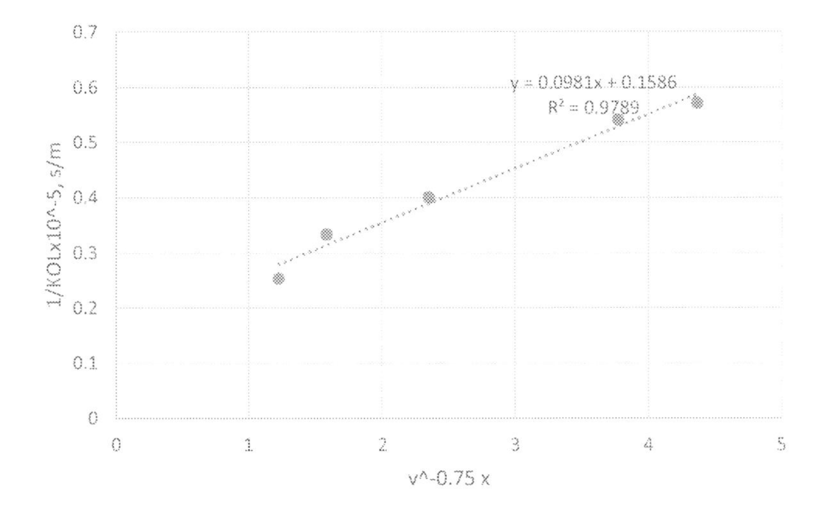

Figure 9.3 Wilson plot with $m = 0.75$.

9.2 Membrane adsorption

9.2.1 Membrane adsorption process outline

Membrane adsorption is a process in which adsorption and membrane filtration are integrated.

Adsorption is a very economic, convenient, and easy operation technique. It is effective especially for heavy metal removal and is applied as a method for all types of wastewater treatments. The special functional groups on the surface of the adsorbents provide significant interactions with heavy metals, resulting in the adsorptive separation of heavy metals from water.

In a membrane adsorbent, functional groups are at the surface and pore wall of polymer membranes and the target pollutants are selectively adsorbed to these functional groups. Thus, the membrane adsorbent effectively combines the filtration and adsorption performance of the membrane. When the contaminated water flows through the membrane, the functional active binding sites will combine with the target contaminants to remove them from drinking water with a high adsorption rate and capacity because of the very short submicron-scale distance from the target pollutants to the active adsorption sites (Fig. 9.4; Khulbe and Matsuura, 2018).

The membrane adsorption experiments are carried out using a continuous filtration setup (Fig. 9.5) with a large reservoir which is connected to a cross-flow permeation cell via a feed supply pump. With the large feed reservoir volume compared to the permeate volume, the feed volume as well as the feed solute

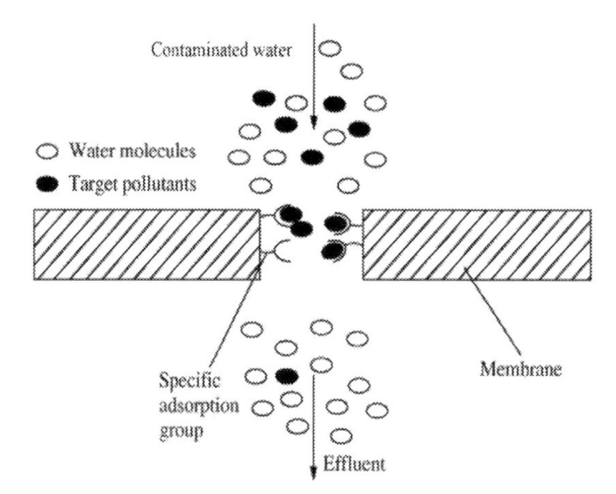

Figure 9.4 The principle of membrane adsorption (Khulbe and Matsuura, 2018).

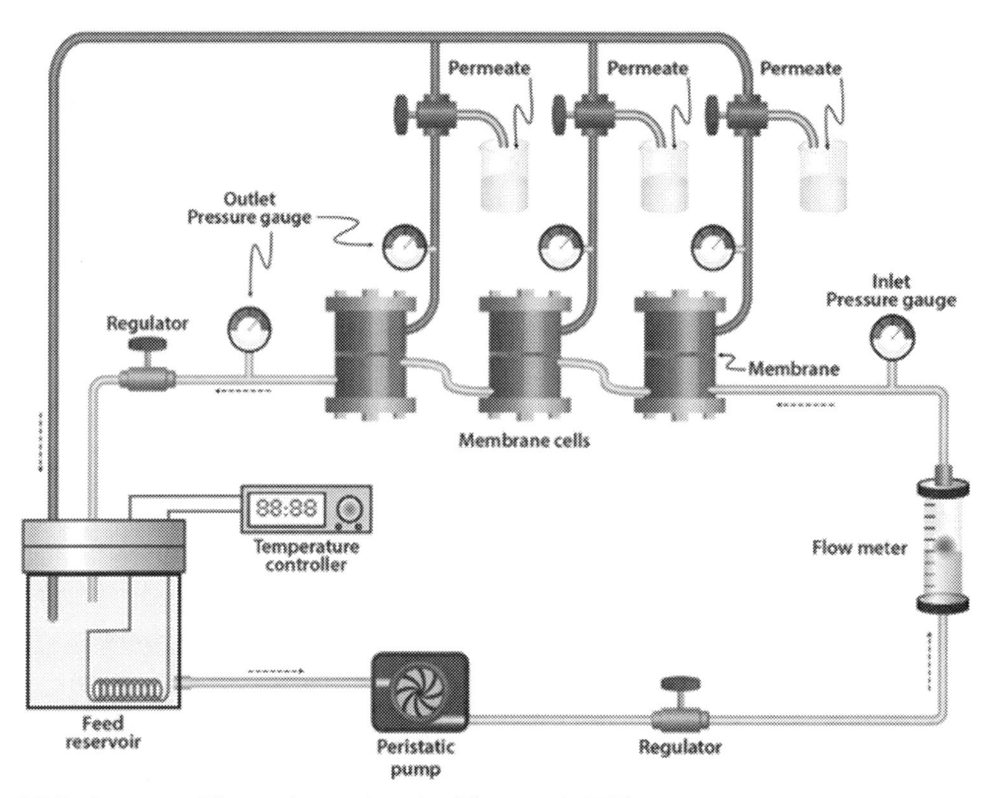

Figure 9.5 Equipment used for membrane adsorption (Efome et al., 2018).

concentration can be assumed to be constant. Permeate is collected from time to time and the permeate flux and solute concentration in the permeate are given as a function of time.

The mathematical model is developed for a nanofibrous membrane adsorbent because of its high throughput compared with the conventional asymmetric flat sheet membrane.

The model consists of two parts. In the first, the Carman–Kozeny (C-K) equation is employed to evaluate the permeate flux of the nanofibrous membrane. The model calculation is made under the condition that the heavy metal concentration is in the ppb range, so that the permeate flux is nearly equal to that of pure water. In the second part, the model is made to reproduce the breakthrough curve of the permeate concentration based on the mass balance equations (Efome et al., 2018).

9.2.2 Carman–Kozeny equation for flux calculation

The membrane flux, J (m^3/m^2 s), is calculated according to Darcy's law using Eq. (9.18).

$$J = \frac{K}{\mu\delta}\Delta p \tag{9.18}$$

where μ is the viscosity of water (Pa s), δ is the membrane thickness (m), and Δp is the transmembrane pressure difference (Pa). K is the permeability coefficient (m^2), which can be calculated by the following C-K equation

$$K = \frac{d_f^2 \varepsilon^3}{16 k_{CK}(1-\varepsilon)^2} \tag{9.19}$$

where ε is the membrane porosity, d_f is the fiber diameter (m). k_{CK} is the C-K constant, a parameter dependent on the structure of the membrane material. For this study, a k_{CK} value of 4.5 is used based on the work of Tomadakis and Robertson (Tomadakis and Robertson, 2005), where they provided k_{CK} for different fiber alignments ranging from randomly oriented fibers to fully aligned fibers.

9.2.3 Mass balance in membrane adsorption

In mass balance, the rate of heavy metal ion outflow from the membrane is the rate of heavy metal ion inflow into the membrane minus the rate of heavy metal adsorption (see Fig. 9.6.) as shown in Eq. (9.20).

$$JAc_p = JAc_f - w\frac{dq_t}{dt} \tag{9.20}$$

where J is water flux (m^3/m^2 s), A is effective membrane area (m^2), c_p and c_f are permeate and feed heavy metal ion concentration (kg/m^3), respectively, w is the mass of the adsorbent embedded membrane (kg), and q_t is the amount of the heavy metal ions adsorbed by the unit mass of the membrane (kg/kg) at time t (s).

As for adsorption, using the first-order kinetics,

$$\frac{dq_t}{dt} = k_1(q_{max} - q_t) \tag{9.21}$$

where q_{max} is the maximum adsorption capacity of the membrane (kg/kg) and k_1 is the pseudo-first-order kinetic constant (s^{-1}).

Integrating,

$$q_t = q_{max}(1 - e^{-k_1 t}) \tag{9.22}$$

From Eqs. (9.21) and (9.22)

$$\frac{dq_t}{dt} = k_1 q_{max} e^{-k_1 t} \tag{9.23}$$

From Eqs. (9.20) and (9.23)

$$JAc_p = JAc_f - wk_1 q_{max} e^{-k_1 t} \tag{9.24}$$

In order to use Eq. (9.24), the following two cases should be considered.

$$\text{(I) } JAc_f < wk_1 q_{max} e^{-k_1 t} \tag{9.25}$$

In this case, the rate of heavy metal ion inflow is less than the rate of adsorption. Then, c_p is zero.

$$\text{(II) } JAc_f \geq wk_1 q_{max} e^{-k_1 t} \tag{9.26}$$

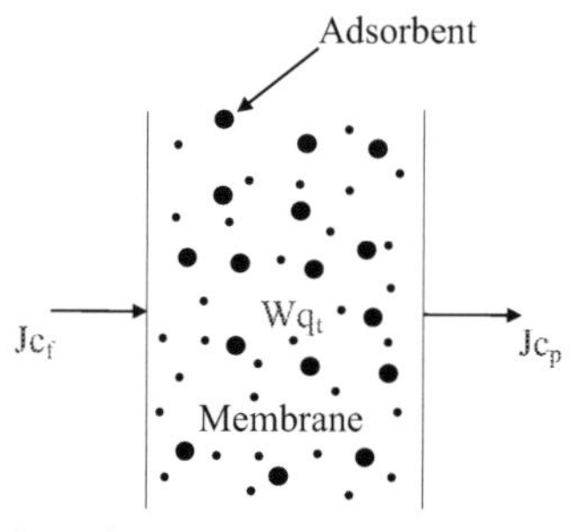

Figure 9.6 Mass balance in membrane adsorption.

In this case, the rate of heavy metal ion influx is more than the rate of adsorption and the heavy metal ion appears in the permeate, making Eq. (9.24) relevant.

Rearranging Eq. (9.24)

$$\ln\left(1-\frac{c_p}{c_f}\right)=\ln\left(\frac{wk_1q_{max}}{JAc_f}\right)-k_1t \tag{9.27}$$

According to Eq. (9.27), k_1 and $\ln\left(\frac{wk_1q_{max}}{JAc_f}\right)$ can be obtained from the slope and the intercept with the y axis of the linear plot, $\ln\left(1-\frac{c_p}{c_f}\right)$ versus t, respectively.

When adsorption and solute rejection by the sieving effect occur simultaneously, representing the sieving effect by the rejection R [defined as $(1-c_p/c_f)$ when there is no adsorption], Eqs. (9.24) and (9.27) become

$$JAc_p=JA(1-R)c_f-wk_1q_{max}e^{-k_1t} \tag{9.28}$$

and

$$\ln\left(1-\frac{c_p}{(1-R)c_f}\right)=\ln\left(\frac{wk_1q_{max}}{JA(1-R)c_f}\right)-k_1t \tag{9.29}$$

respectively.

Using the model developed above for solute flux in the permeate, c_p, can be calculated according to Scheme 9.1.

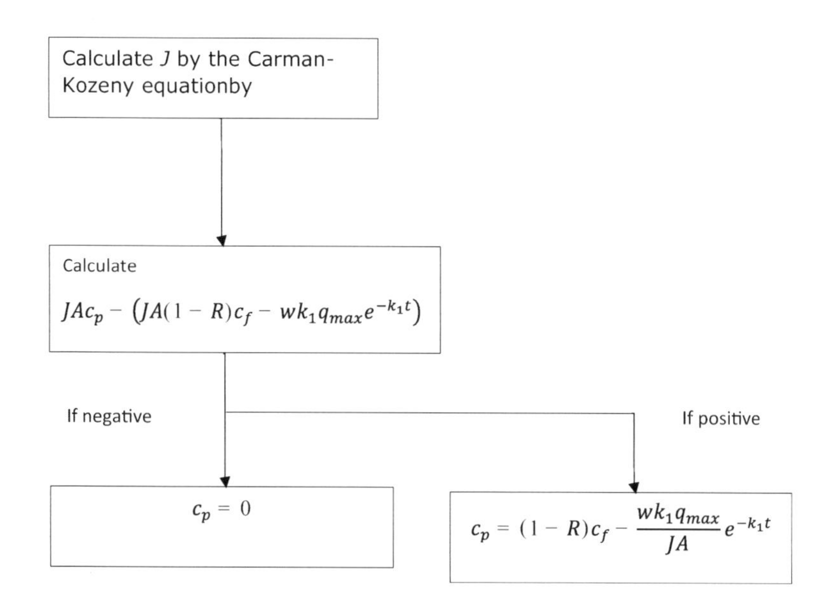

Scheme 9.1 Calculation algorithm for section 9.2.3.

Problem 9.3:

Calculate the membrane flux when the transmembrane pressure difference is 10^4 Pa by using the following parameters:

$$\varepsilon = 0.49$$

$$d_f = 2.027 \times 10^{-7}\text{m}$$

$$k_{CK} = 4.5$$

$$\delta = 3.0 \times 10^{-5}\text{m}$$

and

$$\mu = 8.9 \times 10^{-4} \text{ Pa s}$$

Answer:

From Eqs. (9.18) and (9.19),

$$J = \frac{d_f^2 \varepsilon^3 \Delta p}{16 k_{CK} (1-\varepsilon)^2 \mu \delta}$$
$$= \frac{(2.027 \times 10^{-7})^2 \times 0.49^3 \times 10^4}{16 \times 4.5 \times (1-0.49)^2 \times 8.9 \times 10^{-4} \times 3.0 \times 10^{-5}}$$
$$= 9.667 \times 10^{-5}\text{m}^3/\text{m}^2\text{s}$$

Problem 9.4:

Draw the permeate concentration, c_p, versus time, t, using the following parameters

$$c_f = 25 \times 10^{-6} \frac{\text{kg}}{\text{m}^3} = 25 \text{ mg/L}$$

$$w = 10^{-4}\text{kg} = 100 \text{ mg}$$

$$k_1 = 0.3833 \times 10^{-3} \frac{1}{s} = 1.38 \text{ 1/h}$$

$$q_{max} = 23.98 \times 10^{-3} \frac{\text{kg}}{\text{kg}} = 23.98 \text{ mg/g}$$

$$J = 9.667 \times 10^{-5} \text{ m}^3/\text{m}^2\text{s}$$

$$A = 0.0038 \text{ m}^2$$

and

$$R = 0 \text{ and } 0.3$$

Answer:

For example, for $R = 0$ and $t = 1.2 \times 10^4$ s
$JAc_f = 9.667 \times 10^{-5} \times 0.0038 \times 25 \times 10^{-6} = 9.184 \times 10^{-12}$

$$wk_1 q_{max} e^{-k_1 t} = 10^{-4} \times 0.3833 \times 10^{-3} \times 23.98$$

$$\times 10^{-3} e^{-0.3833 \times 10^{-3} \times 1.2 \times 10^4} = 9.243 \times 10^{-12}$$

$$JAc_f < wk_1 q_{max} e^{-k_1 t}$$

Therefore, $c_p = 0$
For $R = 0$ and $t = 1.5 \times 10^4$ s
$JAc_f = 9.667 \times 10^{-5} \times 0.0038 \times 25 \times 10^{-6} = 9.184 \times 10^{-12}$

$$wk_1 q_{max} e^{-k_1 t} = 10^{-4} \times 0.3833 \times 10^{-3} \times 23.98$$

$$\times 10^{-3} e^{-0.3833 \times 10^{-3} \times 1.5x10^4} = 2.927 \times 10^{-12}$$

$$JAc_f > wk_1 q_{max} e^{-k_1 t}$$

Then,

$$JAc_p = JAc_f - wk_1 q_{max} e^{-k_1 t}$$

$$c_p = c_f - \frac{wk_1 q_{max}}{JA} e^{-k_1 t}$$

$$= 25 \times 10^{-6} - \frac{10^{-4} \times 0.3833 \times 10^{-3} \times 23.98 \times 10^{-3}}{9.667 \times 10^{-5} \times 0.0038}$$

$$\times e^{-0.3833 \times 10^{-3} \times 1.5 \times 10^4} = 17.0 \times 10^{-6}$$

All results are summarized in Table 9.3 and Fig. 9.7.

Table 9.3 The results of the calculation.

$t \times 10^{-4}$, S	$c_p \times 10^6$, kg/m^3	
	$R = 0$	$R = 0.3$
1.2	0	0
1.3	7.85	0.35
1.5	17.0	9.53
1.7	21.3	13.8
2.0	23.8	16.3
2.2	24.5	17.0
2.5	24.8	17.3
3.0	25.0	17.5

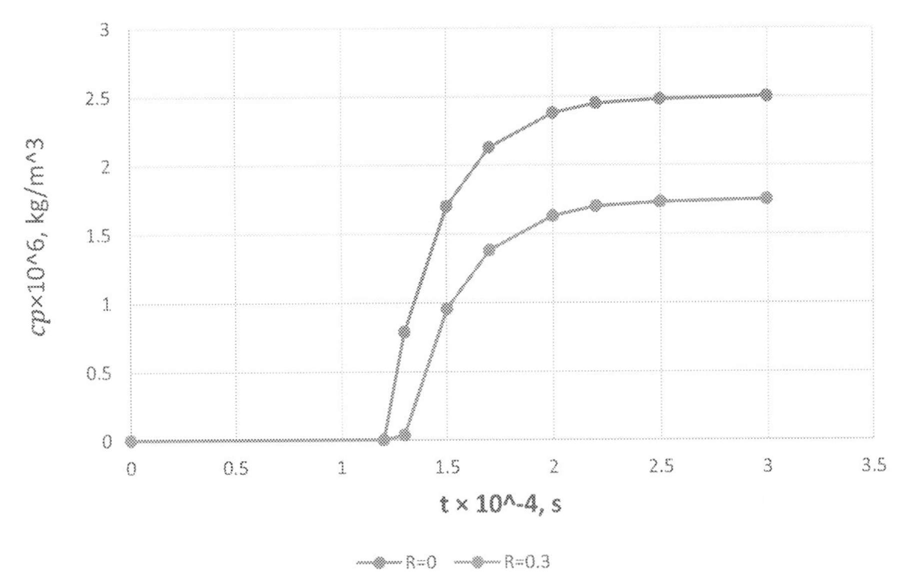

Figure 9.7 c_p versus t.

Nomenclature

Symbol	Definition [unit (SI system)]
For section 9.1	
$\boldsymbol{A}$	Contact area (m^2)
C_g^{in}	CO_2 pressure at hollow fiber inlet (Pa)
C_g^{out}	CO_2 pressure at hollow fiber outlet (Pa)
C_l^{in}	CO_2 concentration in liquid at hollow fiber inlet (mol/m^3)
C_l^{Out}	CO_2 concentration in liquid at hollow fiber outlet (mol/m^3)
$\boldsymbol{d_i}$	Inner diameter of hollow fiber (m)
$\boldsymbol{d_m}$	Log mean diameter of hollow fiber (m)
$\boldsymbol{d_o}$	Outer diameter of hollow fiber (m)
D	Diffusivity of gas in the liquid (m^2/s)
Gz	Graetz number
$\boldsymbol{H}$	Henry's constant (mol/m^3 Pa)
$\boldsymbol{J_{av}}$	Average gas flux (mol/m^2 s)
$\boldsymbol{k_g}$	Gas side mass transfer coefficient (m/s)
$\boldsymbol{k_L}$	Liquid side mass transfer coefficient (m/s)
$\boldsymbol{k_m}$	Membrane mass transfer coefficient (m/s)
$\boldsymbol{K_{OL}}$	Overall mass transfer coefficient (m/s)
$\boldsymbol{L}$	Length of hollow fiber (m)
$\boldsymbol{m}$	Distribution factor, 0.85 in Eq. (9.4)
$\boldsymbol{m}$	Quantity defined by Eq. (9.18)
$\boldsymbol{Q_L}$	Liquid flow rate (m^3/s)
Re	Reynolds number
Sc	Schmidt number
$\boldsymbol{Sh}$	Sherwood number

v_L Liquid velocity (m/s)

Greek letters

ΔC_l^{av} Logarithmic mean of transmembrane concentration difference (mol/m^3)

ρ Density of liquid (kg/m^3)

μ Viscosity of liquid (Pa s)

For section 9.2

A Effective membrane area (m^2)

c_f Solute concentration in feed (kg/m^3)

c_p Solute concentration in permeate (kg/m^3)

d_f Fiber diameter (m)

J Membrane flux (m^3/m^2 s)

k_1 Pseudo-first-order kinetic constant (1/s)

k_{CK} Carman–Kozeny constant (=4.5)

K Permeability coefficient (m^2)

$\boldsymbol{P}$ Pressure (Pa)

q_{max} Maximum adsorption capacity of membrane (kg/kg)

$\boldsymbol{q_t}$ Amount of solute adsorbed by the unit mass of membrane (kg/kg)

$\boldsymbol{R}$ Solute rejection

$\boldsymbol{t}$ Time (s)

$\boldsymbol{W}$ Mass of adsorbent embedded membrane (kg)

Greek letters

δ Membrane thickness (m)

ε Porosity

μ Viscosity (Pa s)

References

Bakeri, G., Ismail, A.F., Shariati, M., Matsuura, T., 2010. Effect of polymer concentration on the structure and performance of polyetherimide hollow fiber membranes using in membrane gas-liquid contacting process. J. Membr. Sci. 363, 103–111.

Bakeri, G., Matsuura, T., Ismail, A.F., Rana, D., 2012. A novel surface modified polyetherimide hollow fiber membrane for gas-liquid contacting processes. Sep. Purif. Technol. 89, 160–170.

Dindore, V.Y., Brilman, D.W.F., Feron, P.H.M., Versteg, G.F., 2004. CO_2 absorption at elevated pressures using a hollow fiber membrane contactor. J. Membr. Sci. 235, 99–109.

Efome, J.E., Rana, D., Matsuura, T., Lan, C.Q., 2018. Experiment and modeling for flux and permeate concentration of heavy metal ion in adsorptive membrane filtration using a metal-organic framework incorporated nanofibrous membrane. Chem. Eng. J 352, 737–744.

Khulbe, K.C., Matsuura, T., 2018. Removal of heavy metals and pollutants by membrane adsorption techniques. Appl. Water Sci. 8, 1–30. article 19.

Kreulen, H., Smolders, C.A., Versteeg, G.F., van Swaaij, W.P.M., 1993. Microporous hollow fiber membrane modules as gas–liquid contactors. Part 1: physical mass transfer processes. J. Membr. Sci. 78, 197–216.

Kumar, P.S., Hogendoorn, J.A., Feron, P.H.M., Versteeg, G.F., 2003. Approximate solution to predict the enhancement factor for the reactive absorption of a gas in a liquid flowing through a microporous membrane hollow fiber. J. Membr. Sci. 213, 231–245.

Li, J.L., Chen, B.H., 2005. Review of CO_2 absorption using chemical solvents in hollow fiber membrane contactors. Sep. Purif. Technol. 41, 109–122.

Tomadakis, M.M., Robertson, T.J., 2005. Viscous permeability of random fiber structures: comparison of electrical and diffusional estimates with experimental and analytical results. J. Compos. Mater. 39, 163–188.

10

Membrane module

10.1 Reverse osmosis

10.1.1 Reverse osmosis hollow fiber module

There are several equations that have been developed for the design of reverse osmosis (RO) hollow fiber modules. Among those, the equations developed by Gill and Bansal (1973), Dandavati et al. (1975), and Gill et al. (1988) are the most rigorous but complicated, while those developed by Ohya and Sourirajan (1969, 1971), Ohya and Taniguchi (1975), and Ohya et al. (1977) are simple but do not allow the calculation of pressure change in the hollow fiber. Taniguchi (1978), Darwish et al. (1989), and Abdel-Jawad and Darwish (1989) basically followed Ohya's approach. Rautenbach's approach includes all three important aspects of hollow fiber modules (velocity, solute concentration, and pressure) and the equations involved are relatively simple (Rautenbach and Dahm, 1987). Therefore his approach is adopted in this section.

Rautenbach considered a bundle of hollow fibers as a uniform body, although it consists of a number of individual fibers. A distribution tube is at the center of the cylinder and the feed solution is distributed evenly in the fiber bundle in the radial direction (see Fig. 10.1; Matsuura, 1994). Each hollow fiber is a

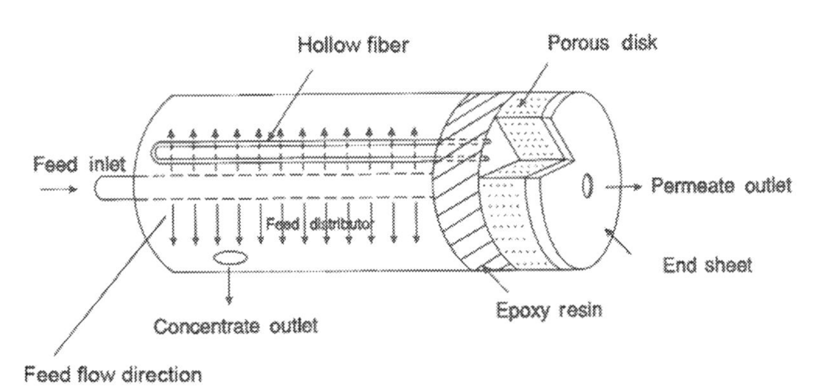

Figure 10.1 Schematic diagram of hollow fiber module (Matsuura, 1994).

Membrane Separation Processes. DOI: https://doi.org/10.1016/B978-0-12-819626-7.00009-0

U-tube; thus one hollow fiber bent in the middle is regarded as two fibers with a closed end. Permeate flows in the lumen side to the longitudinal direction and comes out from the other open end (see Fig. 10.2).

In the model the transport equations in the hollow fiber and those in the hollow fiber bundle were derived separately (Rautenbach and Dahm, 1987). Herein, only the model for a single hollow fiber is given.

Referring to Fig. 10.3, the permeate flow rates into and away from a small segment between the distance x from the dead

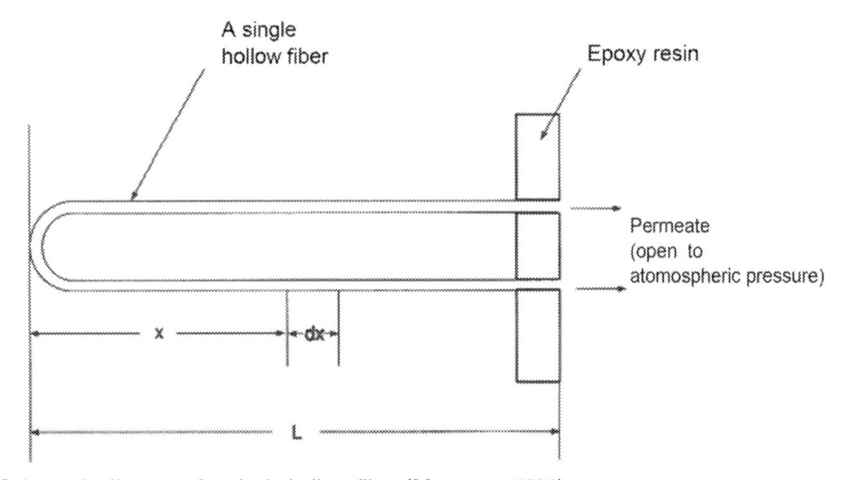

Figure 10.2 Schematic diagram of a single hollow fiber (Matsuura, 1994).

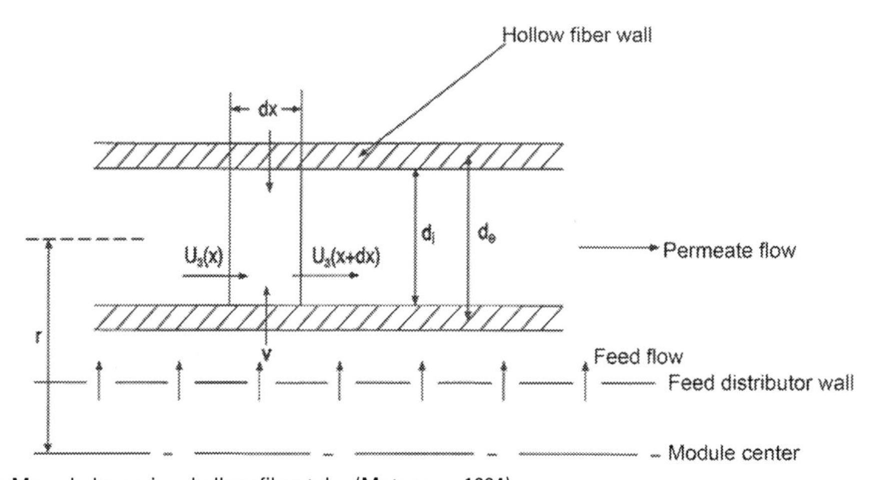

Figure 10.3 Mass balance in a hollow fiber tube (Matsuura, 1994).

end of the hollow fiber (see Fig. 10.2) and $x + dx$ are given, respectively, by

$$\text{Inlet flow rate} = \frac{\pi}{4} d_i^2 u_3 + \pi d_e dx v_w$$

$$\text{Outlet flow rate} = \frac{\pi}{4} d_i^2 \left(u_3 + \frac{du_3}{dx} dx \right)$$

At the steady state, the inlet and the outlet flow rates are balanced, therefore

$$\frac{\pi}{4} d_i^2 u_3 + \pi d_e dx v_w = \frac{\pi}{4} d_i^2 \left(u_3 + \frac{du_3}{dx} dx \right) \tag{10.1}$$

where d_i and d_e are the inside diameter and effective diameter of a hollow fiber tube (m), u_3 is the permeate flow velocity, to the longitudinal direction, in the bore of the hollow fiber tube (m), and v_w is the permeation velocity (m/s) of solvent (water in most cases) based on the effective diameter.

Rearranging,

$$\frac{\pi}{4} d_i^2 \frac{du_3}{dx} = \pi d_e v_w \tag{10.2}$$

Therefore

$$\frac{du_3}{dx} = 4 \frac{d_e}{d_i^2} v_w \tag{10.3}$$

On the other hand,

$$\text{The inlet mass flow rate of the solute} = \frac{\pi}{4} d_i^2 (u_3 \bar{c}_{B3}) + \pi d_e dx J_B$$

where $\bar{c}_{B3}$ (mol/m^3) is the solute concentration on the bore side of the hollow fiber at the longitudinal distance x and J_B (mol/m^2 s) is the solute flux through the hollow fiber membrane.

$$\text{The outlet mass flow rate of the solute} = \frac{\pi}{4} d_i^2 \left((u_3 \bar{c}_{B3}) + \frac{d(u_3 \bar{c}_{B3})}{dx} dx \right)$$

At the steady state

$$\frac{\pi}{4} d_i^2 (u_3 \bar{c}_{B3}) + \pi d_e dx J_B = \frac{\pi}{4} d_i^2 \left((u_3 \bar{c}_{B3}) + \frac{d(u_3 \bar{c}_{B3})}{dx} dx \right) \tag{10.4}$$

Rearranging,

$$\frac{\pi}{4} d_i^2 \frac{d(u_3 \bar{c}_{B3})}{dx} = \pi d_e J_B \tag{10.5}$$

Therefore

$$\frac{d(u_3\bar{c}_{B3})}{dx} = 4\frac{d_e}{d_i^2}J_B \quad (10.6)$$

and

$$\frac{du_3}{dx}\bar{c}_{B3} + u_3\frac{d\bar{c}_{B3}}{dx} = 4\frac{d_e}{d_i^2}J_B \quad (10.7)$$

Combining Eqs. (10.3) and (10.7)

$$4\frac{d_e}{d_i^2}v_w\bar{c}_{B3} + u_3\frac{d\bar{c}_{B3}}{dx} = 4\frac{d_e}{d_i^2}J_B \quad (10.8)$$

Rearranging,

$$\frac{d\bar{c}_{B3}}{dx} = 4\frac{d_e}{u_3 d_i^2}(J_B - v_w\bar{c}_{B3}) \quad (10.9)$$

The pressure drop that occurs while the solution passes through the distance segment dx is given by

$$\frac{dp_3}{dx} = -\frac{32}{d_i^2}\eta u_3 \quad (10.10)$$

The molar solvent flux, J_A (mol/m^2 s), molar solute flux (mol/m^2 s), and the concentration polarization occurring at the feed boundary layer can be written (see Eqs. 3.17, 3.18, and 3.15) as

$$J_A = A\{(p_2 - p_3) - b(c_{B2} - c_{B3})\} \quad (10.11)$$

$$J_B = B(c_{B2} - c_{B3}) \quad (10.12)$$

$$\frac{c_{B2} - c_{B3}}{c_{B1} - c_{B3}} = \exp\left(\frac{v_w}{k}\right) \quad (10.13)$$

Furthermore,

$$v_w = \frac{J_A}{c} \quad (10.14)$$

and

$$c_{B3} = \frac{J_B}{v_w} \quad (10.15)$$

where c is the total molar concentration (mol/m^3) of the permeate solution, which is nearly equal to that of solvent (water), particularly in the desalination process. It is assumed in Eq. (10.11) that the osmotic pressure is proportional to the molar concentration of the solution.

Some explanation is in order for the difference between c_{B3} and $\overline{c}_{B3}$. c_{B3} is the concentration of the permeate through the hollow fiber wall of the small segment whose length is dx. This permeate joins the flow coming from the preceding segment of the hollow fiber, and flows into the next segment as the solution whose concentration is $\overline{c}_{B3}$. c_{B3} and $\overline{c}_{B3}$ are therefore different.

The differential Eqs. (10.3), (10.9), and (10.10) together with the membrane transport Eqs. (10.11)–(10.13) can be solved when a set of data for the feed solution which is on the shell side of the hollow fiber, that is, c_{B1} and p_2 ($p_2 = p_1$ since, unlike concentration, there is no pressure polarization), are given under the following boundary conditions:

$$u_3(x=0)=0 \tag{10.16}$$

$$\overline{c}_{B3}(x=0)=c_{B3}(0) \tag{10.17}$$

and

$$p_3(L)=101.325\ \text{kPa}(=\text{atmospheric pressure}) \tag{10.18}$$

Note that $\overline{c}_{B3}$ and c_{B3} are equal at $x=0$, since there is no preceding segment.

The steps to solve the equations are as follows.

Step 1: The first guess of $p_3(0)$ is made [(0) means $(x=0)$ hereafter].

Step 2: $u_3(\Delta x)$, $\overline{c}_{B3}(\Delta x)$, and $p_3(\Delta x)$ are obtained. [(Δx) means $(x=\Delta x)$].

Step 2.1: $J_A(0), J_B(0), v_w(0), c_{B2}(0), c_{B3}(0)$ are obtained by solving Eqs. (10.11)–(10.15) simultaneously.

Step 2.2: The derivatives $(\frac{du_3}{dx})_{x=0}$, $(\frac{d\overline{c}_{B3}}{dx})_{x=0}$, and $(\frac{dp_3}{dx})_{x=0}$ are obtained.

From Eq. (10.3),

$$\left(\frac{du_3}{dx}\right)_{x=0}=4\frac{d_e}{d_i^2}v_w(0) \tag{10.19}$$

From Eqs. (10.10) and (10.16),

$$\left(\frac{dp_3}{dx}\right)_{x=0}=0 \tag{10.20}$$

As for $(\frac{d\overline{c}_{B3}}{dx})_{x=0}$, Eqs. (10.16) and (10.17) as the boundary conditions make the right side of Eq. (10.9) indeterminate at $x=0$, since both the numerator and denominator are zero. Hence, the following L'Hopital's rule is applied.

$$\left(\frac{d\overline{c}_{B3}}{dx}\right)_{x=0}=4\frac{d_e}{d_i^2}\frac{\{d(J_B-v_w\overline{c}_{B3})/dx\}_{x=0}}{(du_3/dx)_{x=0}} \tag{10.21}$$

Combining Eqs. (10.19) and (10.21),

$$\left(\frac{d\overline{c}_{B3}}{dx}\right)_{x=0} = \frac{1}{v_w(0)}\{d(J_B - v_w\overline{c}_{B3})/dx\}_{x=0} \tag{10.22}$$

and

$$(d\overline{c}_{B3})_{x=0} = \frac{1}{v_w(0)}\{d(J_B - v_w\overline{c}_{B3})\}_{x=0} \tag{10.23}$$

Then,

$$\overline{c}_{B3}(dx) - \overline{c}_{B3}(0) = \frac{1}{v_w(0)}[\{J_B(dx) - v_w(dx)\overline{c}_{B3}(dx)\} - \{J_B(0) - v_w(0)\overline{c}_{B3}(0)\}] \tag{10.24}$$

From Eqs. (10.15) and (10.17)

$$J_B(0) - v_w(0)\overline{c}_{B3}(0) = 0$$

Therefore from Eq. (10.24)

$$\overline{c}_{B3}(dx) = \overline{c}_{B3}(0) + \frac{1}{v_w(0)}\{J_B(dx) - v_w(dx)\overline{c}_{B3}(dx)\} \tag{10.25}$$

Rearranging

$$\overline{c}_{B3}(dx)\left\{1 + \frac{v_w(dx)}{v_w(0)}\right\} = \overline{c}_{B3}(0) + \frac{J_B(dx)}{v_w(0)} \tag{10.26}$$

and

$$\overline{c}_{B3}(dx) = \frac{\overline{c}_{B3}(0) + \frac{J_B(dx)}{v_w(0)}}{1 + \frac{v_w(dx)}{v_w(0)}} \tag{10.27}$$

For a small increment Δx

$$\overline{c}_{B3}(\Delta x) = \frac{\overline{c}_{B3}(0) + \frac{J_B(\Delta x)}{v_w(0)}}{1 + \frac{v_w(\Delta x)}{v_w(0)}} \tag{10.28}$$

Eq. (10.28) will be used in the next step to know $\overline{c}_{B3}(\Delta x)$.

Step 2.3: $u_3(\Delta x), \overline{c}_{B3}(\Delta x)$, and $p_3(\Delta x)$ are obtained for a preset size of the increment Δx.

$$u_3(\Delta x) = u_3(0) + \left(\frac{du_3}{dx}\right)_{x=0} \Delta x \tag{10.29}$$

where Eq. (10.19) can be used.

$$p_3(\Delta x) = p_3(0) + \left(\frac{dp_3}{dx}\right)_{x=0} \Delta x = p_3(0) \tag{10.30}$$

Eq. (10.30) was derived from Eq. (10.20).

When Eqs. (10.11)–(10.15) are solved for $x = \Delta x$, the answers will be the same as at $x = 0$, since

$$p_3(\Delta x) = p_3(0).$$

Then, again using Eqs. (10.15) and (10.17)

$$\frac{J_B(\Delta x)}{v_w(0)} = \frac{J_B(0)}{v_w(0)} = c_{B3}(0) = \bar{c}_{B3}(0)$$

$$\frac{v_w(\Delta x)}{v_w(0)} = \frac{v_w(0)}{v_w(0)} = 1$$

Therefore from Eq. (10.28)

$$\bar{c}_{B3}(\Delta x) = \bar{c}_{B3}(0) \tag{10.31}$$

Step 3: $u_3(2\Delta x)$, $\bar{c}_{B3}(2\Delta x)$, and $p(2\Delta x)$ are obtained.

Step 3.1: $J_A(\Delta x), J_B(\Delta x), v_w(\Delta x), c_{B2}(\Delta x)$, and $c_{B3}(\Delta x)$ have already been obtained in Step 2.3, when Eqs. (10.11)–(10.15) were solved at $x = \Delta x$.

Step 3.2: Using those values, $(\frac{du_3}{dx})_{x=\Delta x}$, $(\frac{d\bar{c}_{B3}}{dx})_{x=\Delta x}$, and $(\frac{dp_3}{dx})_{x=\Delta x}$ are calculated.

Step 3.3: $u_3(2\Delta x)$, $\bar{c}_{B3}(2\Delta x)$, and $p_3(2\Delta x)$ are then obtained by

$$u_3(2\Delta x) = u_3(\Delta x) + \left(\frac{du_3}{dx}\right)_{x=\Delta x} \Delta x \tag{10.32}$$

$$\bar{c}_{B3}(2\Delta x) = \bar{c}_{B3}(\Delta x) + \left(\frac{d\bar{c}_{B3}}{dx}\right)_{x=\Delta x} \Delta x \tag{10.33}$$

$$p_3(2\Delta x) = p_3(\Delta x) + \left(\frac{dp_3}{dx}\right)_{x=\Delta x} \Delta x \tag{10.34}$$

Step 4: The above steps are repeated until $x = L$ is reached. It is checked if $p_3(L)$ is 101.325 kPa. If it is, then the computation is stopped. Otherwise, we go back to Step 1 and a new $p_3(0)$ is guessed.

[The method to solve Eqs. (10.11)–(10.15) for a given set of parameters, $A, B, b, p_2, p_3, c_{B2}, k$, and c, is as follows.

1. Assume $c_{B2} - c_{B3}$.
2. Calculate J_A and J_B from Eqs. (10.11) and (10.12).
3. Calculate v_w from Eq. (10.14).
4. Calculate c_{B3} from Eq. (10.15)
5. Check if $c_{B2} - c_{B3}, c_{B3}$, and v_w can satisfy Eq. (10.14).

If satisfied, they are the answers. If not, go back to (1) and continue until Eq. (10.14) is satisfied.]

The simulation algorithm is shown in Scheme 10.1.

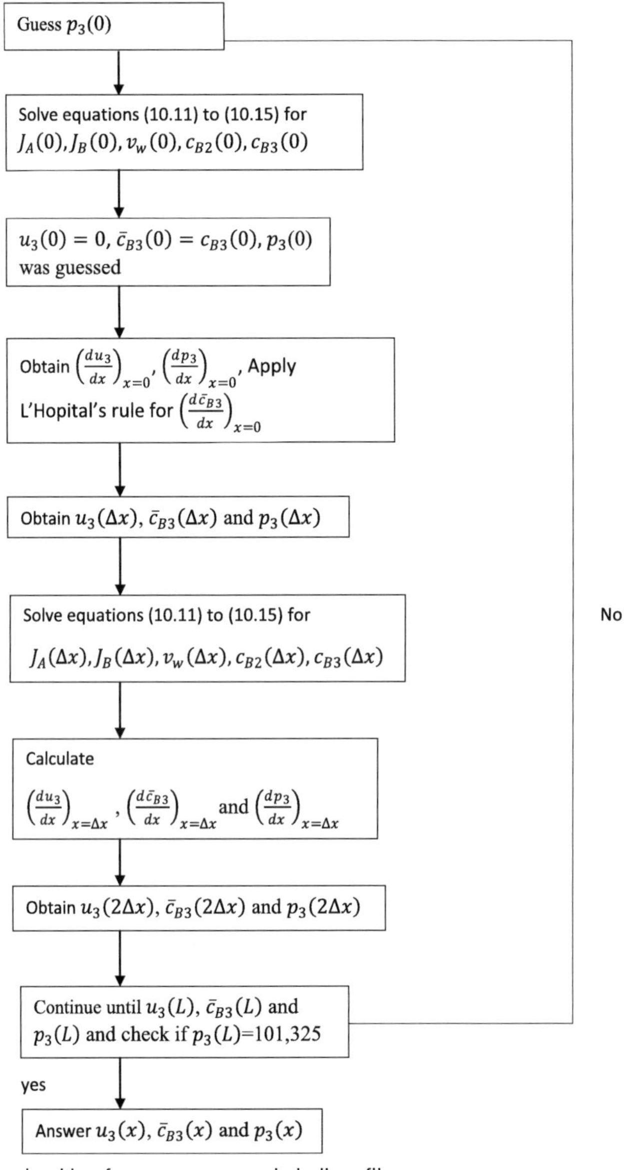

Scheme 10.1 Calculation algorithm for reverse osmosis hollow fiber.

Problem 10.1:

Calculate the longitudinal velocity, u_3, solute concentration, $\bar{c}_{B3}$, and pressure, p_3, in the lumen side of hollow fiber as functions of the distance, x, from the closed end of the hollow fiber assuming no concentration polarization on the shell side of the membrane. Use the following parameters.

Concentration of sodium chloride on the shell side: $c_{B1} = 0.1 \text{ kmol/m}^3$

Pressure on the shell side of the membrane: $p_1 = p_2 = 2758 \text{ kPa (gage)}$

Pure water permeation coefficient: $A = 0.2 \times 10^{-7} \text{ kmol/m}^2 \text{ s kPa}$

Solute permeation constant: $B = 0.075 \times 10^{-7} \text{ m/s}$

Total molar concentration: $c = 55.3 \text{ kmol/m}^3$

Length of hollow fiber: $L = 0.6 \text{ m}$

Effective hollow fiber diameter: $d_e = 0.000105 \text{ m}$

Inner diameter of hollow fiber: $d_i = 0.00010 \text{ m}$

Viscosity of water at 25°C: $\eta = 0.8941 \times 10^{-6} \text{kPa s}$

Osmotic pressure coefficient: $b = 4.6374 \times 10^3 \text{ kPa m}^3\text{/kmol}$

Answer:

Step 1: The pressure in the hollow fiber lumen at the sealed end ($x = 0$ in Fig. 10.2) is assumed to be 120 kPa (absolute).

Step 2:

Step 2.1: $J_A(0), J_B(0), c_{B3}(0)$, and v_w (0) are obtained by solving Eqs. (10.11)–(10.15).

Since concentration polarization can be ignored the mass transfer coefficient, $k = \infty$. Therefore from Eq. (10.13), $c_{B2} = c_{B1}$.

From Eqs. (10.14) and (10.15)

$$c_{B3}(0) = \frac{J_B(0)c}{J_A(0)} \tag{10.35}$$

Solving Eqs. (10.11), (10.12), and (10.35) simultaneously with $c_{B2} = c_{B1}$, we obtain

$$J_A(0) = \frac{-\{Bc - A(p_2 - p_3(0)) + Abc_{B1}\} + \sqrt{\{Bc - A(p_2 - p_3(0)) + Abc_{B1}\}^2 + 4A(p_2 - p_3(0))Bc}}{2} \tag{10.36}$$

$$J_B(0) = \frac{Bc_{B1}}{1 + \frac{Bc}{J_A(0)}} \tag{10.37}$$

Inserting numerical values into Eq. (10.36)

$$\begin{aligned}\alpha &= Bc - A(p_2 - p_3(0)) + Abc_{B1}\\ &= 0.075 \times 10^{-7} \times 55.3 - 0.2 \times 10^{-7} \times (2758 + 101.3 - 120)\\ &\quad + 0.2 \times 10^{-7} \times 4.64 \times 10^3 \times 0.1\\ &= -450.9 \times 10^{-7}\end{aligned}$$

$$J_A(0) = \frac{-450.9 + \sqrt{(-450.9)^2 + 4 \times 0.2 \times 10^{-7} \times (2758 + 101.3 - 120) \times 0.075 \times 10^{-7} \times 55.3}}{2}$$

$$= 456 \times 10^{-7} \text{kmol/m}^2\text{s}$$

From Eq. (10.37)

$$J_B(0) = \frac{0.075 \times 10^{-7} \times 0.1}{1 + \frac{0.075 \times 10^{-7} \times 55.3}{456 \times 10^{-7}}} = 0.00743 \times 10^{-7} \ \text{kmol/m}^2\ \text{s}$$

From Eq. (10.35)

$$c_{B3}(0) = \frac{0.00743 \times 10^{-7} \times 55.3}{456 \times 10^{-7}} = 9.02 \times 10^{-4} \ \text{kmol/m}^3$$

From Eq. (10.14)

$$v_w(0) = \frac{456 \times 10^{-7}}{55.3} = 8.24 \times 10^{-7} \ \text{m/s}$$

Step 2.2: The derivative $\left(\frac{du_3}{dx}\right)_{x=0}$ is obtained.
From Eq. (10.19)

$$\left(\frac{du_3}{dx}\right)_{x=0} = 4\frac{0.000105}{(0.000100)^2} \times 8.24 \times 10^{-7} = 34.6 \times 10^{-3}$$

Step 2.3: $u_3(\Delta x), \overline{c}_{B3}(\Delta x)$, and $p_3(\Delta x)$ are obtained for a preset size of the increment $\Delta x = 0.1$.

($\Delta x = 0.1$ may be too large. But this is used only to explain how the problem is solved. Much smaller Δx can be used to obtain much more precise answers.)

Using the boundary condition (10.16),

$$u_3(0.1) = u_3(0) + 34.6 \times 10^{-3} \times 0.1 = 3.46 \times 10^{-3}$$

From boundary the condition (10.17) and (10.31)

$$\bar{c}_{B3}(0.1) = \bar{c}_{B3}(0) = c_{B3}(0) = 9.02 \times 10^{-4}$$

And from Eq. (10.30) and the guessed value,

$$p_3(0.1) = p_3(0) = 120$$

Step 3: $u_3(0.2)$, $\bar{c}_{B3}(0.2)$, and $p_3(0.2)$ are obtained.

Step 3.1: $J_A(0.1)$, $J_B(0.1)$, $c_{B3}(0.1)$, and v_w (0.1) are the same as $J_A(0)$, $J_B(0)$, $c_{B3}(0)$, and v_w (0).

Step 3.2: Using those values, $(\frac{du_3}{dx})_{x=0.1}$, $(\frac{d\bar{c}_{B3}}{dx})_{x=0.1}$, and $(\frac{dp_3}{dx})_{x=0.1}$ are obtained by Eqs. (10.3), (10.9), and (10.10), respectively.

$$\left(\frac{du_3}{dx}\right)_{x=0.1} = 4\frac{0.000105}{(0.000100)^2} \times 8.24 \times 10^{-7} = 34.6 \times 10^{-3}$$

$$\begin{aligned}\left(\frac{d\bar{c}_{B3}}{dx}\right)_{x=0.1} &= \frac{4 \times 0.000105}{3.46 \times 10^{-4} \times 0.0001^2} \\ &\quad \times \left(0.00743 \times 10^{-7} - 8.24 \times 10^{-7} \times 9.02 \times 10^{-4}\right) \\ &= -3 \times 10^{-5}\end{aligned}$$

$$\left(\frac{dp_3}{dx}\right)_{x=0.1} = -\frac{32}{0.000100^2} \times 0.8941 \times 10^{-6} \times 3.46 \times 10^{-3} = -9.90$$

Step 3.3: $u_3(0.2)$, $\bar{c}_{B3}(0.2)$, and $p_3(0.2)$ are obtained.

$$u_3(0.2) = 3.46 \times 10^{-3} + 34.6 \times 10^{-3} \times 0.1 = 6.92 \times 10^{-3}$$

$$\bar{c}_{B3}(0.2) = 9.02 \times 10^{-4} - 3 \times 10^{-5} \times 0.1 = 8.99 \times 10^{-4}$$

$$p_3(0.2) = 120 - 9.90 \times 0.1 = 119$$

$$\vdots$$

$$u_3(0.6) = 20.8 \times 10^{-3} \text{ m/s}$$

$$\bar{c}_{B3}(0.6) = 8.94 \times 10^{-4} \text{ kmol/m}^3$$

$$p_3(0.6) = 105 > 101.2 \text{ kPa}$$

...

Therefore the initial guess of $p_3(0) = 120$ kPa is not a good choice.

An initial guess of 116 kPa satisfies the boundary condition $p_3(0.6) = 101$ kPa.

The changes of $u_3(x)$, $\bar{c}_{B3}(x)$, and $p_3(x)$ are shown in Fig. 10.4.

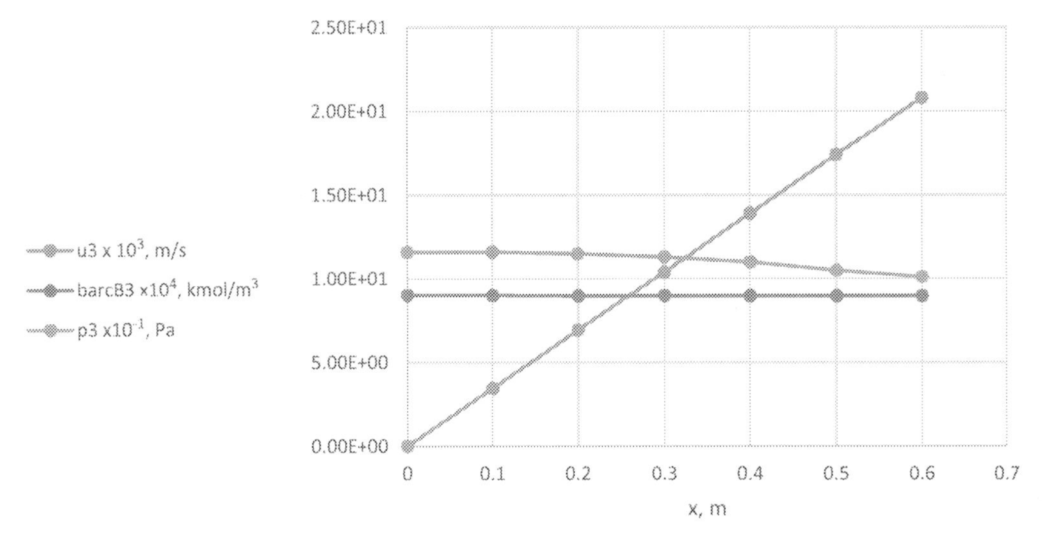

Figure 10.4 $u_3(x)$, $\bar{c}_{B3}(x)$, and $p_3(x)$ versus x.

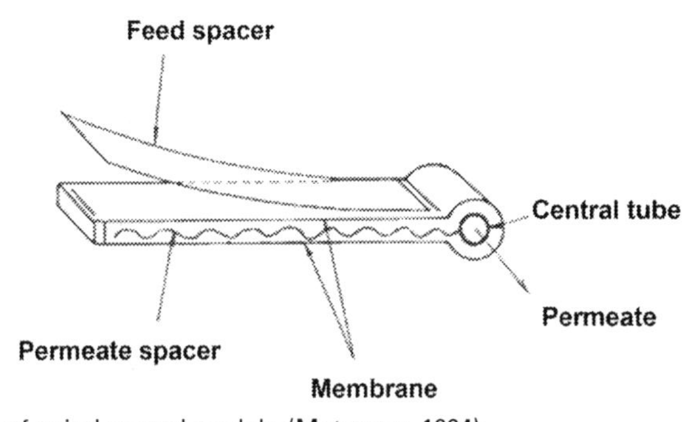

Figure 10.5 Basic structure of spiral-wound module (Matsuura, 1994).

10.1.2 Reverse osmosis flat-sheet membrane

The spiral-wound module is the most popular module for RO featuring high-pressure durability, compactness, minimum membrane contamination, minimum concentration polarization, and minimum pressure drop in the permeate channel. Its basic structure is illustrated schematically in Fig. 10.5. A permeate spacer is sandwiched between two membranes. The porous support side of the membrane faces the permeate spacer. Three edges of the membranes are sealed with glue to form a membrane envelope, of which the open end is connected to a perforated central tube. The membrane envelope so produced is wound spirally around the central tube together with the feed spacer. To make the leaf length shorter, several membrane

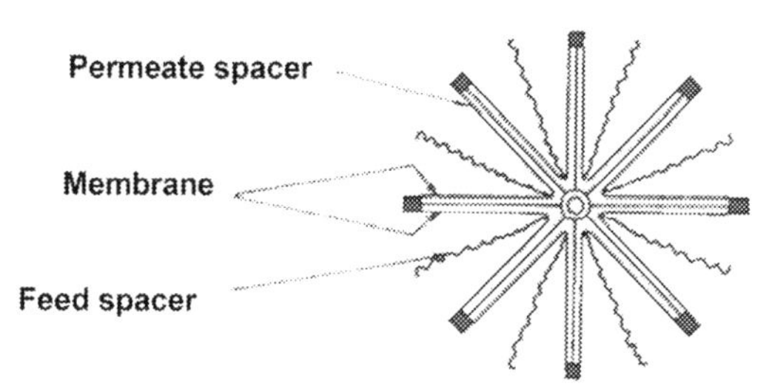

Figure 10.6 Multileaves spiral-wound module (Matsuura, 1994).

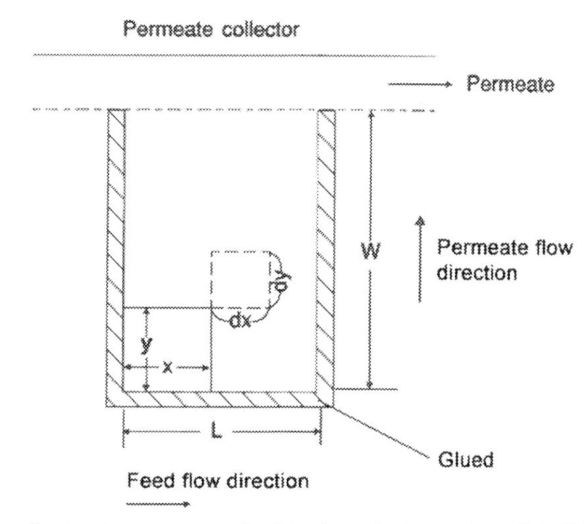

Figure 10.7 Rectangular coordinate of a membrane leaf in the spiral-wound module (Matsuura, 1994).

leaves are wound simultaneously, as illustrated in Fig. 10.6. The feed solution flows through the feed spacer parallel to the central tube, while the permeate flows through the permeate spacer, spirally, perpendicular to the feed flow direction, and is collected into the central tube.

The analysis of the spiral-wound module was attempted by Ohya and Taniguchi (Ohya and Taniguchi, 1975), and later Rautenbach (Rautenbach and Dahm, 1987) elaborated their model.

Fig. 10.7 illustrates the rectangular coordinate for one leaf of spiral-wound membrane. The feed solution flows along the x-axis. It enters at $x=0$ and leaves at $x=L$. The permeate solution flows in the y-direction. The flow starts from the sealed end of the envelope at $y=0$ and leaves the envelope at $y=W$ to enter into the collection tube.

In the model the following assumptions are made.

1. The pressure and concentration of feed solution are uniform in the entire area of the membrane. (This assumption was made to simplify the mathematical equations for demonstration purposes.)
2. Immediate and complete mixing of permeate with the main flow in the permeate channel of the membrane leaf.
3. Plug flow in the permeate channel.
4. Solute transport by diffusion is neglected inside the permeate channel in both the x- and y-directions.
5. Under these assumptions, only the change to parameters in the y-direction is considered.

Considering a small distance segment of dy at a position y,

$$\text{Input flow rate} = \varepsilon_p h_p L u_3(y) + 2\varepsilon_f L dy v_w$$

$$\text{Output flow rate} = \varepsilon_p h_p L u_3(y + dy)$$

where ε_p and ε_f are the void ratio of the permeate channel and feed channel spacer, and h_p is the permeate channel thickness.

At the steady state,

$$\varepsilon_p h_p L u_3(y) + 2\varepsilon_f L dy v_w = \varepsilon_p h_p L u_3(y + dy) \tag{10.38}$$

Since

$$u_3(y + dy) = u_3(y) + \frac{du_3(y)}{dy} dy \tag{10.39}$$

Substituting Eq. (10.39) for $u_3(y + dy)$ in Eq. (10.38) and further rearranging,

$$\frac{du_3(y)}{dy} = \frac{2v_w}{h_p} \frac{\varepsilon_f}{\varepsilon_p} \tag{10.40}$$

$$\text{Solute input} = \varepsilon_p h_p L u_3(y) \bar{c}_{B3}(y) + 2\varepsilon_p L dy J_B \tag{10.41}$$

$$\text{Solute output} = \varepsilon_p h_p L u_3(y) \bar{c}_{B3}(y) + \varepsilon_p h_p L \frac{du_3(y) \bar{c}_{B3}(y)}{dy} dy \tag{10.42}$$

At the steady state,

$$\varepsilon_p h_p L u_3(y) \bar{c}_{B3}(y) + 2\varepsilon_f L dy J_B = \varepsilon_p h_p L u_3(y) \bar{c}_{B3}(y) + \varepsilon_p h_p L \frac{du_3(y) \bar{c}_{B3}(y)}{dy} dy \tag{10.43}$$

Hence,

$$\frac{du_3(y)\bar{c}_{B3}(y)}{dy} = \frac{2J_B}{h_p}\frac{\varepsilon_f}{\varepsilon_p} \tag{10.44}$$

$$\frac{du_3(y)\bar{c}_{B3}(y)}{dy} = \frac{du_3(y)}{dy}\bar{c}_{B3}(y) + u_3(y)\frac{d\bar{c}_{B3}(y)}{dy} \tag{10.45}$$

Using Eq. (10.40)

$$\frac{du_3(y)\bar{c}_{B3}(y)}{dy} = \frac{2v_w}{h_p}\frac{\varepsilon_f}{\varepsilon_p}\bar{c}_{B3}(y) + u_3(y)\frac{d\bar{c}_{B3}(y)}{dy} \tag{10.46}$$

Combining Eqs. (10.44) and (10.46),

$$\frac{d\bar{c}_{B3}(y)}{dy} = \frac{2(J_B - v_w\bar{c}_{B3}(y))}{u_3(y)h_p}\frac{\varepsilon_f}{\varepsilon_p} \tag{10.47}$$

At $y = 0$, L'Hopital's rule will apply and,

$$\frac{d\bar{c}_{B3}(y)}{dy} = \frac{2d(J_B - v_w\bar{c}_{B3}(y))/dy}{du_3(y)/dy}\frac{\varepsilon_f}{h_p\varepsilon_p} \tag{10.48}$$

which leads to

$$\frac{d\bar{c}_{B3}(y)}{dy} = 0 \tag{10.49}$$

[Under the boundary condition $\bar{c}_{B3}(0) = c_{B3}(0)$ and Eq. (10.15).] Therefore Eq. (10.31) is also applicable in this case.

For the pressure drop

$$\frac{dp_3(y)}{dy} = \frac{\rho}{2}u_3(y)^2\frac{\lambda_p}{2h_p} \tag{10.50}$$

where

$$\lambda_p = 1075Re^{0.78} \tag{10.51}$$

$$Re = \frac{\rho u_3(y)h_p}{\eta_{water}} \tag{10.52}$$

and ρ is the density of water.

The same membrane transport Eqs. (10.11)–(10.15) are applicable, that is,

$$J_A = A\{(p_2 - p_3) - b(c_{B2} - c_{B3})\} \tag{10.11}$$

$$J_B = B(c_{B2} - c_{B3}) \tag{10.12}$$

$$\frac{c_{B2}-c_{B3}}{c_{B1}-c_{B3}}=\exp\left(\frac{v_w}{k}\right) \tag{10.13}$$

Furthermore,

$$v_w=\frac{J_A}{c} \tag{10.14}$$

and

$$c_{B3}=\frac{J_B}{v_w} \tag{10.15}$$

Problem 10.2:

Under the similar boundary conditions in Eqs. (10.16)–(10.18), that is,

$$u_3(y=0)=0 \tag{10.16}$$

$$\overline{c}_{B3}(y=0)=c_{B3} \tag{10.17}$$

and

$$p_3(W)=101.325\ \text{kPa}(=\text{atmospheric pressure}) \tag{10.18}$$

Calculate $u_3, \overline{c}_{B3}$, and p_3 as the function of y, using the following parameters. Ignore concentration polarization.

Concentration of sodium chloride on the shell side: $c_{B2}=0.1\ \text{kmol/m}^3$;

Pressure on the shell side of the membrane: $p_1=2758\ \text{kPa (gage)}$;

Pure water permeation coefficient: $A=0.2\times10^{-7}\ \text{kmol/m}^2\,\text{s kPa}$;

Solute permeation constant: $B=0.075\times10^{-7}\ \text{m/s}$;

Total molar concentration: $c=55.3\ \text{kmol/m}^3$;

Length of permeate channel: $W=0.6\ \text{m}$;

Void space of feed channel: $\varepsilon_f=0.9$;

Void space of permeate channel: $\varepsilon_p=0.9$;

$$\text{Permeate channel thickness}: h_p=0.001\ \text{m};$$

$$\text{Density of water at } 25^\circ\text{C}: \rho=997\frac{\text{kg}}{\text{m}^3};$$

$$\text{Viscosity of water at } 25^\circ\text{C}: \eta=0.8941\times10^{-6}\ \text{kPa s};$$

$$\text{Osmotic pressure coefficient}: b=4.6374\times10^3\ \frac{\text{kPa m}^3}{\text{kmol}}.$$

Answer:

Step 1: The pressure in the hollow fiber lumen at the sealed end ($y=0$ in Fig. 10.7) is assumed to be 850 kPa (absolute).

Step 2.1: $J_A(0)$, $J_B(0)$, $c_{B3}(0)$, and $v_w(0)$ are obtained by solving Eqs. (7.11)–(7.15).

Since concentration polarization can be ignored the mass transfer coefficient, $k=\infty$, from Eq. (10.13) $c_{B2}=c_{B1}$.

Eq. (10.36) is applicable in this case.

$$J_A(0) = \frac{-\{Bc - A(p_2 - p_3(0)) + Abc_{B1}\} + \sqrt{\{Bc - A(p_2 - p_3(0)) + Abc_{B1}\}^2 + 4A(p_2 - p_3(0))Bc}}{2} \qquad (10.36)$$

Inserting numerical values into Eq. (10.36)

$$\begin{aligned}\alpha &= Bc - A(p_2 - p_3(0)) + Abc_{B1}\\ &= 0.075\times10^{-7}\times55.3 - 0.2\times10^{-7}\\ &\quad\times(2758+101.3-850) + 0.2\times10^{-7}\times4637\times0.1\\ &= -304.9\times10^{-7}\end{aligned}$$

$$J_A(0) = \frac{304.9\times10^{-7} + \sqrt{(-304.9\times10^{-7})^2 + 4\times0.2\times10^{-7}\times(2758+101.3-850)\times0.075\times10^{-7}\times55.3}}{2}$$

$$= 310.3\times10^{-7}\ \text{kmol/m}^2\,\text{s}$$

From Eqs. (10.37), (10.35), and (10.14),

$$J_B(0) = \frac{0.075\times10^{-7}\times0.1}{1+\frac{0.075\times10^{-7}\times55.3}{310.3\times10^{-7}}} = 0.00763\times10^{-7}\ \text{kmol/m}^2\,\text{s}$$

$$c_{B3}(0) = \frac{0.00763\times10^{-7}\times55.3}{310.3\times10^{-7}} = 1.36\times10^{-3}\ \text{kmol/m}^3$$

$$v_w(0) = \frac{310.3\times10^{-7}}{55.3} = 5.61\times10^{-7}\ \text{m/s}$$

Step 2.2: The derivative $\left(\frac{du_3}{dy}\right)_{y=0}$ is obtained.
From Eq. (10.40)

$$\left(\frac{du_3}{dy}\right)_{y=0} = \frac{2\times5.61\times10^{-7}}{10^{-3}}\times\frac{0.9}{0.9} = 11.2\times10^{-4}$$

Step 2.3: $u_3(\Delta y), \bar{c}_{B3}(\Delta y)$, and $p_3(\Delta y)$ are obtained for a preset size of the increment $\Delta y = 0.1$.

($\Delta y = 0.1$ may be too large. But this was used only to explain how the problem is solved.)

Using the boundary condition (10.16),

$$u_3(0.1) = u_3(0) + 11.2 \times 10^{-4} \times 0.1 = 11.2 \times 10^{-5}$$

From boundary condition (10.17) and (10.31)

$$\overline{c}_{B3}(0.1) = \overline{c}_{B3}(0) = c_{B3}(0) = 1.36 \times 10^{-3}$$

From boundary condition (10.16) and (10.50)

$$p_3(0.1) = p_3(0) = 850$$

Step 3: $u_3(0.2)$, $\overline{c}_{B3}(0.2)$, and $p_3(0.2)$ are obtained.

Step 3.1: $J_A(0.1), J_B(0.1), c_{B3}(0.1)$, and $v_w(0.1)$ are the same as $J_A(0), J_B(0), c_{B3}(0)$, and $v_w(0)$.

Step 3.2: Using those values, $(\frac{du_3}{dy})_{y=0.1}$, $(\frac{d\overline{c}_{B3}}{dx})_{y=0.1}$, and $(\frac{dp_3}{dx})_{y=0.1}$ are obtained.

$$\left(\frac{du_3}{dy}\right)_{y=0.1} = \frac{2 \times 5.61 \times 10^{-7}}{10^{-3}} \times \frac{0.9}{0.9} = 11.2 \times 10^{-4}$$

$$\frac{d\overline{c}_{B3}(y)}{dy} = \frac{2(J_B - v_w \overline{c}_{B3}(y))}{u_3(y) h_p} \frac{\varepsilon_f}{\varepsilon_p} \tag{10.47}$$

$$\left(\frac{d\overline{c}_{B3}}{dy}\right)_{y=0.1} = \frac{2}{11.2 \times 10^{-5} \times 0.001} \times \left(0.00763 \times 10^{-7} - 5.61 \times 10^{-7} \times 1.36 \times 10^{-3}\right) \frac{0.9}{0.9} = 7.143 \times 10^{-7}$$

$$\frac{dp_3(y)}{dy} = \frac{\rho}{2} u_3(y)^2 \frac{\lambda_p}{2h_p} \tag{10.50}$$

where

$$\lambda_p = 1075 Re^{0.78} \tag{10.51}$$

$$Re = \frac{\rho u_3(y) h_p}{\eta_{\text{water}}} = \frac{997 \times 11.2 \times 10^{-5} \times 0.001}{0.8941 \times 10^{-6}} = 124.9 \tag{10.52}$$

$$\left(\frac{dp_3}{dy}\right)_{y=0.1} = -\frac{997}{2} \times \left(11.2 \times 10^{-5}\right)^2 \times \frac{1075 \times 124.9^{0.78}}{2 \times 0.001} = -145.1$$

$$u_3(0.2) = 11.2 \times 10^{-5} + 11.2 \times 10^{-4} \times 0.1 = 2.24 \times 10^{-4}$$

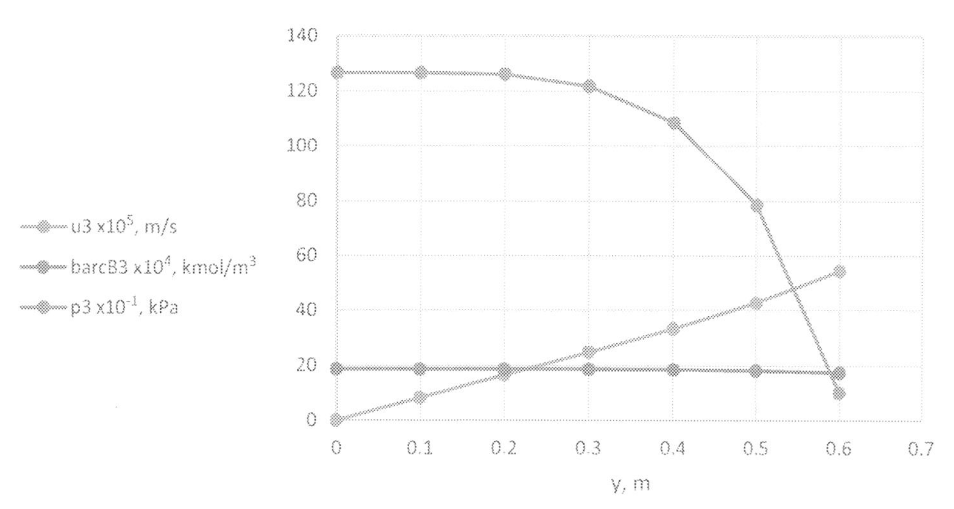

Figure 10.8 $u_3(x)$, $\overline{c}_{B3}(x)$, and $p_3(x)$ versus y.

$$\overline{c}_{B3}(0.2) = 1.36 \times 10^{-3} - 7.143 \times 10^{-7} \times 0.1 = 1.36 \times 10^{-3}$$

$$p_3(0.2) = 850 - 145.1 \times 0.1 = 835.5$$

$$\vdots$$

$$p_3(0.6) = -2072 < 101.3 \text{ kPa}$$

The initial guess of $p_3(0) = 1265.7\text{kPa}$ satisfies the boundary condition, $p_3(0.6) = 101.3$ kPa.

The changes to $u_3(y)$, $\overline{c}_{B3}(y)$, and $p_3(y)$ are shown in Fig. 10.8.

10.2 Gas separation

The mathematical model for the separation of a gas mixture, A and B, by a hollow fiber that is operated in a concurrent mode to the feed gas flow was developed by Pan and Habgood (1978a,b). The following assumptions were made in this model.

1. The membrane permeability is independent of the pressure and concentration.
2. The pressure drop in the feed gas is negligible.
3. The pressure drop of the permeate in the feed is governed by the Hagen–Poiseuille law.

Under the above assumptions the following mass and material balance and pressure drop equations were derived. The method to derive the model equations is similar to RO. Fig. 10.9

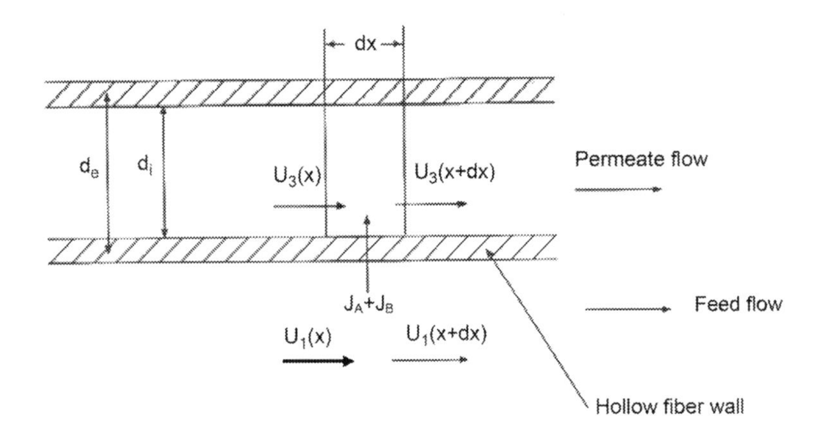

Figure 10.9 Hollow fiber working in concurrent mode (Matsuura, 1994).

illustrates schematically a hollow fiber that is working in concurrent mode.

From the mass balance of the permeate gas:

$$\frac{d\boldsymbol{u}_3(x)}{dx} = \frac{4d_e}{d_i^2}(J_A + J_B) \tag{10.53}$$

where $\boldsymbol{u}_3$ (mol/m^2 s) was used instead of u_3 (m/s) to emphasize the difference of their dimensions. J_A and J_B are the flux (mol/m^2 s) of component A and component B, respectively. Other symbols are the same as those used in the foregoing sections.

From the material balance of the permeate gas:

$$\frac{d\overline{X}_{A3}(x)}{dx} = 0, \quad \text{when } x = 0 \tag{10.54}$$

by L'Hopital's rule.

Otherwise,

$$\frac{d\overline{X}_{A3}(x)}{dx} = \frac{4d_e}{\boldsymbol{u}_3 d_i^2}\{J_A\overline{X}_{B3}(x) - J_B\overline{X}_{A3}(x)\} \tag{10.55}$$

where $\overline{X}_{A3}(x)$ and $\overline{X}_{B3}(x)$ are the mole fraction of components A and B, respectively, in the gas mixture flowing in the permeate channel.

The pressure drop of permeate is:

$$\frac{dp_3(x)^2}{dx} = \frac{64\mu RT}{d_i^2}\boldsymbol{u}_3 \tag{10.56}$$

From the mass balance of the feed and permeate:

$$S_1\boldsymbol{u}_1(x) + S_3\boldsymbol{u}_3(x) = S_1\boldsymbol{u}_1(0) \tag{10.57}$$

$$S_1\boldsymbol{u}_1(x)X_{A1}(x) + S_3\boldsymbol{u}_3(x)\overline{X}_{B3}(x) = S_1\boldsymbol{u}_1(0)X_{A1}(0) \tag{10.58}$$

where S_1 and S_3 are the cross-sectional areas of the lumen side of the hollow fiber and that of the space between the hollow fiber.

The feed pressure is constant, therefore

$$p_1(x) = p_1(0) \quad \text{for } 0 < x \leq L \tag{10.59}$$

For the gas mixture in the feed channel,

$$X_{A1}(x) + X_{B1}(x) = 1 \tag{10.60}$$

For the permeating gas mixture,

$$X_{A3}(x) + X_{B3}(x) = 1 \tag{10.61}$$

For the gas mixture in the permeate channel,

$$\overline{X}_{A3}(x) + \overline{X}_{B3}(x) = 1 \tag{10.62}$$

The transport equations for gas permeation are

$$J_A = \frac{P_A}{\delta}\left(p_1 X_{A1} - p_3 X_{A3}\right) \tag{10.63}$$

$$J_B = \frac{P_B}{\delta}\left(p_1 X_{B1} - p_3 X_{B3}\right) \tag{10.64}$$

$$X_{A3} = \frac{J_A}{J_A + J_B} \tag{10.65}$$

The mass balance Eqs. (10.53) and (10.57), the material balance Eqs. (10.54), (10.55), and (10.58), and equations for pressure (10.56) and (10.59) can be solved with the membrane transport Eqs. (10.63)–(10.65), with the following boundary conditions.

$$\boldsymbol{u}_3(0) = 0 \tag{10.66}$$

$$\overline{X}_{A3}(0) = X_{A3}(0) \tag{10.67}$$

$$p_3(L) = 101.3 \text{ kPa} \tag{10.68}$$

The simulation algorithm is shown in Scheme 10.2.

Problem 10.3:

A CO_2(A)/N_2(B) gas mixture is separated by a hollow fiber membrane made of polyimide 6FDA-DABA.

Calculate $\boldsymbol{u}_3, \overline{X}_{A3}$, and p_3 as a function of x, using the following parameters.

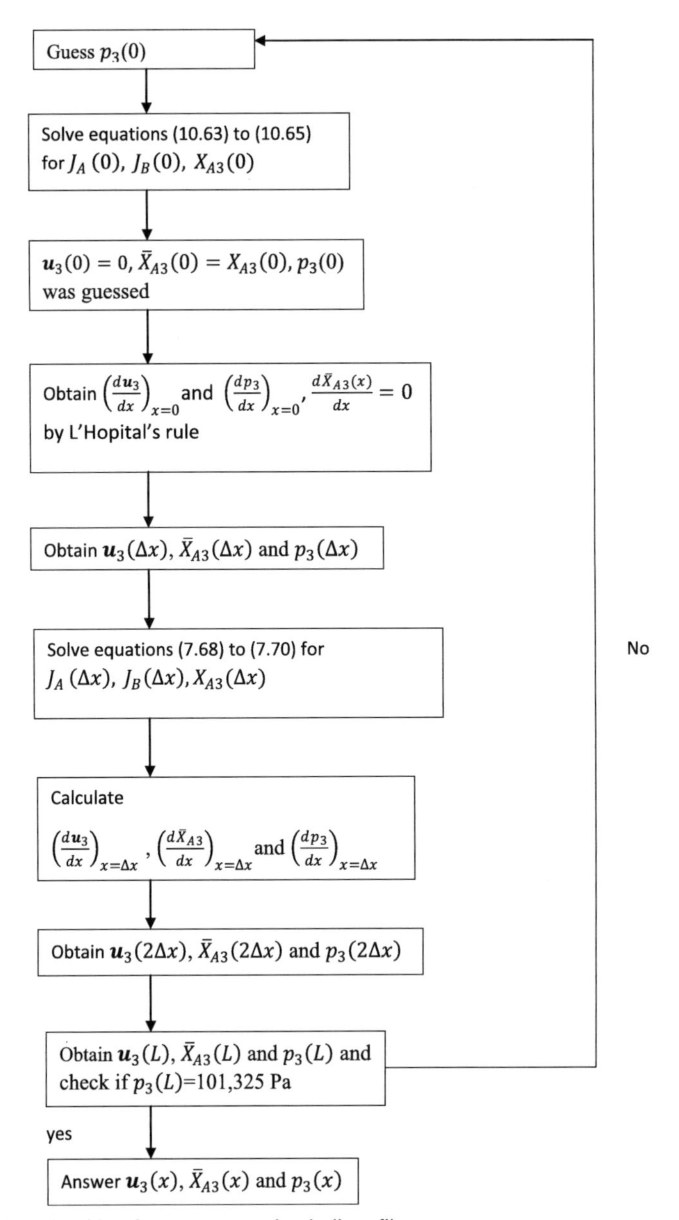

Scheme 10.2 Calculation algorithm for gas separation hollow fiber.

Molar gas velocity of the feed at the sealed end of hollow fiber: $u_1(0) = 200\ \text{mol/m}^2\ \text{s}$;

Feed CO_2 mole fraction of the feed at the sealed end of hollow fiber: $X_{A1}(0) = 0.2$;

Feed gas pressure: $p_1(0) = 10$ atmospheric pressure: 1.01325×10^6 Pa (absolute);

CO_2 permeance: $P_A/\delta = 26.3$ GPU $= 89.2 \times 10^{-10}$ mol/m^2 s Pa;

N_2 permeance: $P_B/\delta = 0.93$ GPU $= 3.15 \times 10^{-10}$ mol/m^2 s Pa;

Effective diameter of hollow fiber: $d_e = 1.05 \times 10^{-4}$ m;

Inner diameter of hollow fiber: $d_i = 1.00 \times 10^{-4}$ m;

Hollow fiber length: $L = 0.6$ m;

Viscosity of CO_2/N_2 gas mixture: $\mu = 1.663 \times 10^{-5}$Pa s;

The ratio of the cross-sectional area of the shell side to that of the lumen side is 1.0;

Temperature: 35°C.

Answer:

Step 1: The pressure in the hollow fiber lumen at the sealed end $p_3(0)$ is assumed to be 1.2×10^5 Pa.

Step 2

Step 2.1: $J_A(0), J_B(0), X_{A3}(0)$ are obtained by solving Eqs. (10.63)–(10.65).

From Eqs. (6.30) and (6.31) of Chapter 6, Membrane Gas Separation,

$$X_{A3} = \frac{-\left(P_A p_1 X_{A1} + P_B p_1 X_{B1} + P_A p_3 - P_B p_3\right) + \alpha}{2(P_B p_3 - P_A p_3)} \tag{10.69}$$

where

$$\alpha = \sqrt{\left(P_A p_1 X_{A1} + P_B p_1 X_{B1} + P_A p_3 - P_B p_3\right)^2 + 4 \times (P_B p_3 - P_A p_3) \times P_A p_1 X_{A1}} \tag{10.70}$$

Using the given numerical values and solving,

$$\alpha = 0.001454$$

$$X_{A3}(0) = 0.794$$

From Eq. (10.63)

$$J_A(0) = 89.2 \times 10^{-10} \times \left(1.01325 \times 10^6 \times 0.2 - 1.2 \times 10^5 \times 0.794\right) = 9.57 \times 10^{-4}$$

From Eq. (10.64)

$$J_B(0) = 3.15 \times 10^{-10} \times \left(1.01325 \times 10^6 \times 0.8 - 1.2 \times 10^5 \times 0.206\right) = 2.48 \times 10^{-4}$$

Thus

$$\overline{X}_{A3}(0) = X_{A3}(0) = 0.794$$

Step 2.2: The derivative $(\frac{du_3}{dx})_{x=0}$ is obtained.

From Eq. (10.53)

$$\frac{d\boldsymbol{u}_3(0)}{dx} = \frac{4 \times 1.05 \times 10^{-4}}{\left(1.00 \times 10^{-4}\right)^2}\left(9.57 \times 10^{-4} + 2.48 \times 10^{-4}\right) = 50.6$$

From Eq. (10.54)

$$\frac{d\overline{X}_{A3}(0)}{dx} = 0$$

From Eq. (10.55)

$$\frac{dp_3(0)}{dx} = 0$$

Step 2.3: $\boldsymbol{u}_3(\Delta x), \overline{X}_{A3}(\Delta x)$, and $p_3(\Delta x)$ are obtained for a preset size of the increment $\Delta x = 0.1$.

($\Delta x = 0.1$ may be too large. But this was used only to explain how the problem is solved.)

$$\boldsymbol{u}_3(0.1) = 0 + 50.6 \times 0.1 = 5.06$$

$$\overline{X}_{A3}(0.1) = 0.794 + 0 = 0.794$$

$$p_3(0.1) = 1.2 \times 10^5 + 0 = 1.2 \times 10^5$$

Step 3: $\boldsymbol{u}_3(0.2)$, $\overline{X}_{A3}(0.2)$, and $p_3(0.2)$ are obtained.

Step 3.1: $J_A(0.1), J_B(0.1), X_{A3}(0.1)$ are obtained by solving Eqs. (10.58), (10.63), (10.64), (10.69), and (10.70).

$$\text{With } X_{A1}(0.1) = \frac{\left(\boldsymbol{u}_1(0)X_{A1}(0) - \boldsymbol{u}_3(0.1)\overline{X}_{A3}(0.1)\right)}{\left(\boldsymbol{u}_1(0) - \boldsymbol{u}_3(0.1)\right)} = \frac{(200 \times 0.2 - 5.06 \times 0.794)}{200 - 5.06} = 0.185$$

$$\alpha = \sqrt{\left(P_A p_1 X_{A1} + P_B p_1 X_{B1} + P_A p_3 - P_B p_3\right)^2 + 4 \times (P_B p_3 - P_A p_3) \times P_A p_1 X_{A1}}$$

$$\alpha = 0.00137$$

$$X_{A3}(0.1) = 0.770$$

$$J_A(0.1) = 8.43 \times 10^{-4}$$

$$J_B(0.1) = 2.52 \times 10^{-4}$$

From Eqs. (10.53)

$$\frac{d\boldsymbol{u}_3(0.1)}{dx} = \frac{4 \times 1.05 \times 10^{-4}}{\left(1.00 \times 10^{-4}\right)^2}\left(8.43 \times 10^{-4} + 2.52 \times 10^{-4}\right) = 46.0$$

From Eqs. (10.55)

$$\frac{d\overline{X}_{A3}(0.1)}{dx} = \frac{4 \times 1.05 \times 10^{-4}}{5.06 \times \left(1.00 \times 10^{-4}\right)^2}\left\{8.43 \times 10^{-4} \times 0.206 - 2.52 \times 10^{-4} \times 0.794\right\}$$
$$= -0.22$$

From Eqs. (10.56)

$$\frac{dp_3(x)}{dx} = \frac{64\mu RT}{2p_3 d_i^2}\boldsymbol{u}_3$$

$$\frac{dp_3(0.1)}{dx} = -\frac{64 \times 1.663 \times 10^{-5} \times 8.314 \times 308.2}{2 \times 1.2 \times 10^5 \times \left(1.00 \times 10^{-4}\right)^2} \times 5.06 = -5739$$

$$\boldsymbol{u}_3(0.2) = 5.06 + 46.0 \times 0.1 = 9.66$$

$$\overline{X}_{A3}(0.2) = 0.794 - 0.22 \times 0.1 = 0.772$$

$$p_3(0.2) = 1.2 \times 10^5 - 5739 \times 0.1 = 1.194 \times 10^5$$

.
.
.

$$p_3(0.6) = 1.12 \times 10^5 \text{ Pa} > 101.3 \text{ kPa}$$

Initial guess of $p_3(0) = 1.10 \times 10^5$ Pa satisfies the boundary condition of $p_3(0.6) = 101.3$ kPa.

Finally, we obtain

$$\boldsymbol{u}_3(0.6) = 25.8 \text{ mol/m}^2 \text{ s}$$

$$\overline{X}_{A3}(0.6) = 0.737$$

$$p_3(0.6) = 101.3 \text{ kPa}$$

The changes of $\boldsymbol{u}_3(x)$, $\overline{X}_{A3}(x)$, and $p_3(x)$ are shown in Fig. 10.10.

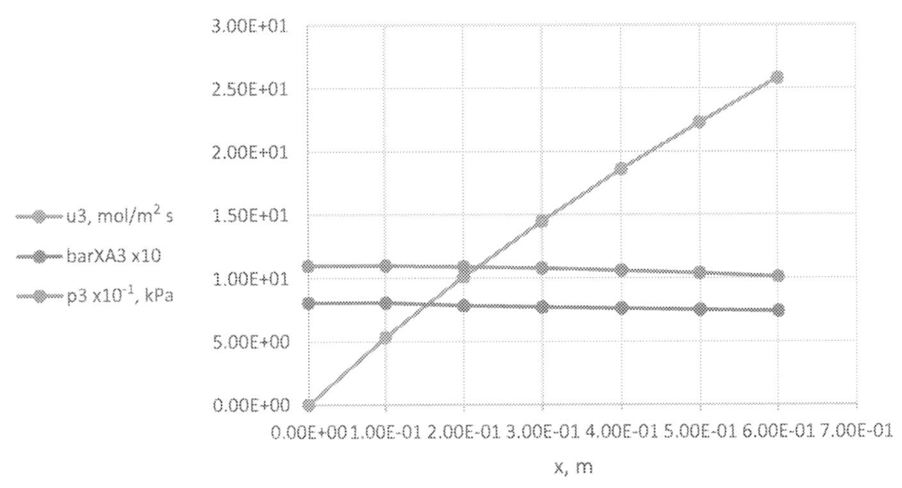

Figure 10.10 $u_3(x)$, $\overline{X}_{A3}(x)$, and $p_3(x)$ versus x.

Nomenclature

Symbols	Definition [dimension (in SI units)]
A	Pure water permeation coefficient (mol/m^2 s Pa)
b	Osmotic pressure coefficient (Pa m^3/mol)
B	Solute permeation constant (m/s)
c	Total molar concentration, nearly equal to concentration of solvent (water) (mol/m^3)
c_{B1}	Solute concentration in the bulk feed solution (mol/m^3)
c_{B2}	Solute concentration in the concentrated boundary layer (mol/m^3)
c_{B3}	Solute concentration in the solution permeating through membrane (mol/m^3)
$\overline{c}_{B3}$	Solute concentration in the lumen side of hollow fiber, or in the permeate channel (mol/m^3)
d_e	Effective diameter of hollow fiber (m)
d_i	Inner diameter of hollow fiber (m)
h_p	Thickness of permeate flow channel (m)
J_A	Solvent flux or flux of component A (mol/m^2 s)
J_B	Solute flux or flux of component B (mol/m^2 s)
k	Mass transfer coefficient (m/s)
L	Length of hollow fiber or width of spiral-wound leaf (m)
p_1	Pressure on the feed side (Pa)
p_2	$=p_1$ (Pa)
p_3	Pressure on the permeate side (Pa)
P_A	Permeability for component A (mol m/m^2 s Pa)
P_B	Permeability for component B (mol m/m^2 s Pa)
Re	Reynolds number
S_1	Total cross-sectional area of the space between hollow fibers (m^2)
S_3	Total lumen cross-sectional area of follow fibers (m^2)
u_3	Permeate flow velocity in the lumen side of hollow fiber, or in the permeate spacer in case of spiral-wound module (m/s)

u_1	Molar flow rate of feed gas mixture per unit cross-sectional area of the space between hollow fibers (mol/m^2 s)
u_3	Molar flow rate of permeate per unit area of hollow fiber cross-section (mol/m^2 s)
v_w	Permeation velocity of solvent (water) (m/s)
W	Length of spiral-wound permeate channel (m)
x	Distance from the sealed end of hollow fiber (m)
X_{A1}	Mole fraction of component A in feed gas mixture
X_{A3}	Mole fraction of component A in the gas mixture permeating through the membrane
X_{B1}	Mole fraction of component B in feed gas mixture
X_{B3}	Mole fraction of component B in the gas mixture permeating through the membrane
$\overline{X}_{A3}$	Mole fraction of component A flowing on the lumen side of membrane
$\overline{X}_{B3}$	Mole fraction of component B flowing on the lumen side of membrane
y	Distance from the sealed end of spiral-wound leaf (m)

Greek letters

δ	Active layer thickness of membrane (m)
ε_p	Void ratio of permeate flow channel
η	Viscosity (Pa s)
λ_p	Quantity defined by Eq. (10.51)
μ	Viscosity of gas mixture (Pa s)
ρ	Solution density (kg/m^3)

References

Abdel-Jawad, M., Darwish, M.A., 1989. Simplified analysis of transport in reverse osmosis (RO) hollow fibers (HF) membranes. Desalination 75, 97–116.

Dandavati, M.S., Doshi, M.R., Gill, W.N., 1975. Hollow fiber of reverse osmosis: experiments and analysis of radial flow systems. Chem. Eng. Sci. 30, 877.

Darwish, B.A.Q., Aly, G.S., Al-Rqobah, H.A., Abdel-Jawad, M., 1989. Predictability of membrane performance of reverse osmosis systems for desalination. Desalination 75, 55–69.

Gill, W.N., Bansal, B., 1973. Hollow fiber reverse osmosis systems analysis and design. AIChE J. 823, 137–152.

Gill, W.N., Matsumoto, M.R., Gill, A.L., Lee, Y.-T., 1988. Flow patterns in radial flow hollow fiber reverse osmosis systems. Desalination 68, 11–28.

Matsuura, T., 1994. Synthetic Membranes and Membrane Separation Processes. CRC Press, Boca Raton, FL, Chapter 7.

Ohya, H., Sourirajan, S., 1969. Effect of longitudinal diffusion in reverse osmosis. AIChE J. 15, 829.

Ohya, H., Sourirajan, S., 1971. Reverse Osmosis System Specification and Performance Data for Water Treatment. Thayer School of Engineering, Dartmouth College, Dartmouth, NH.

Ohya, H., Taniguchi, Y., 1975. An analysis of reverse osmotic characteristics of ROGA-4000 spiral-wound module. Desalination 16, 359–373.

Ohya, H., Nakajima, N., Takagi, K., Kagawa, S., Negishi, Y., 1977. An analysis of reverse osmotic characteristics of B-9 hollow fiber module. Desalination 21, 257–274.

Pan, C.-Y., Habgood, H.W., 1978a. Gas separation by permeation Part I. Calculation methods and parametric analysis. Can. J. Chem. Eng. 56, 197–209.

Pan, C.-Y., Habgood, H.W., 1978b. Gas separation by permeation Part II. Effect of permeate pressure drop and choice of permeate pressure. Can. J. Chem. Eng. 56, 210–217.
Rautenbach, R., Dahm, W., 1987. Design and optimization of spiral-wound and hollow fiber RO-modules. Desalination 65, 259–275.
Taniguchi, Y., 1978. An analysis of reverse osmosis characteristics of ROGA spiral-wound modules. Desalination 25, 71–88.

11

Membrane system

11.1 Two flow types

There are many types of system design in membrane separation processes. Among those, the most appropriate has to be chosen to minimize the cost to achieve the required quality and quantity of the product.

Regarding the membrane module, there are two basic flow types. One is dead end and the other is cross flow. In the dead end type, the entire feed fluid is forced to permeate the membrane. This flow type is found in old filtration systems but is still used in some microfiltration processes. In most industrial applications the cross flow type is used. The deposit on the membrane surface is swept away while the feed fluid flows at the membrane surface in the lateral direction, making the membrane fouling less severe. The permeate passes through the membrane perpendicular to the feed flow. These flow types are shown schematically in Fig. 11.1.

11.1.1 Cross-flow types

In analogy to chemical reactor design, the cross-flow type consists of complete mixing and plug flow. In complete mixing the feed fluid that flows into a vessel is completely mixed before it comes out of the vessel as the retentate. Thus, the composition in the vessel is the same as that of the retentate.

The plug flow consists of co-current and counter-current flow. In co-current flow the feed and permeate flow in the same direction and in counter-current flow they flow in opposite directions. The feed and permeate composition change from the feed entrance to the feed exit and the composition of the retentate is that of the feed at the feed exit. The order in the separation efficiency is as follows:

counter-current flow > co-current flow > complete mixing.

The cross flow types are schematically illustrated in Fig. 11.2.

Membrane Separation Processes. DOI: https://doi.org/10.1016/B978-0-12-819626-7.00003-X

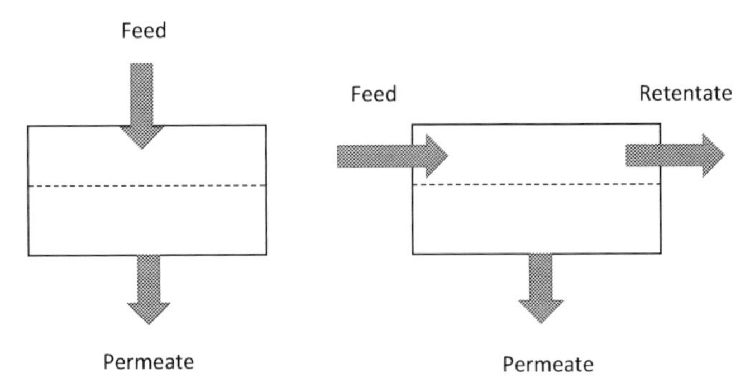

Figure 11.1 Two flow types (left, dead end; right, cross flow).

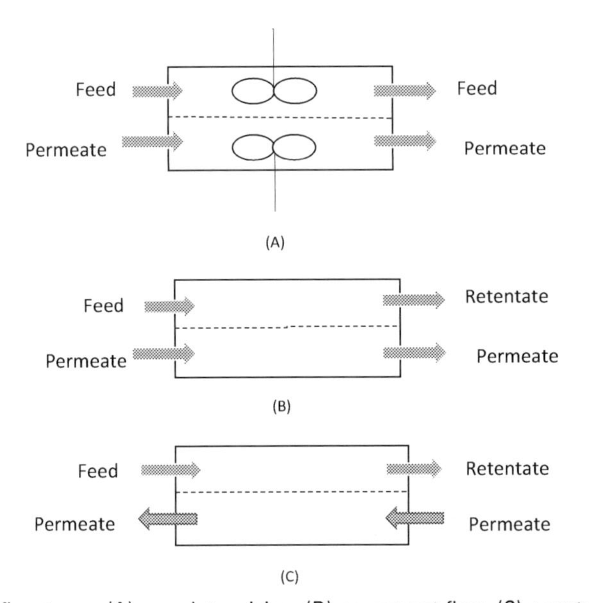

Figure 11.2 Various cross flow types: (A) complete mixing, (B) co-current flow, (C) counter-current flow.

11.1.2 Cascade and recycle

To increase the separation efficiency, the membrane module is operated either in cascade or recycle.

In cascade, one of the outflows, either permeate or retentate, of the first-stage module is supplied to the second stage as the feed. The choice of permeate or retentate depends on which one is the targeted product. The number of stages may be more than two.

In the recycle, a portion of one of the outflows, either permeate or retentate, is recycled back to the feed inlet of the module. Again, the choice of the outflow depends on which one is the targeted product.

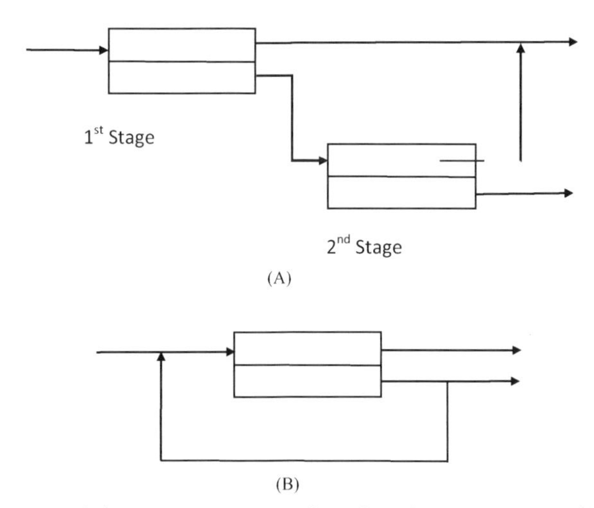

Figure 11.3 Cascade and recycle: (A) cascade, permeate flows into the second stage, (B) recycle, permeate is recycled.

The cascade and recycle are schematically illustrated in Fig. 11.3.

Thus, a number of systems can be designed by combining different flow types.

11.1.3 Hybrid systems

Different membrane processes can be combined to enhance the separation efficiency of the individual processes. For reverse osmosis (RO) alone the following hybrid systems have been proposed and studied:

Reverse osmosis—distillation;
Microfiltration—reverse osmosis;
Ultrafiltration—reverse osmosis;
Nanofiltration—reverse osmosis;
Forward osmosis—reverse osmosis;
Pressure-retarded osmosis—reverse osmosis;
Pervaporation—reverse osmosis;
Reverse osmosis—electrodialysis;
Reverse osmosis—electrodialysis reversal;
Reverse osmosis—ion exchange.

There are a number of other possibilities, including other membrane separation processes.

The membrane processes are often combined with a chemical reactor to facilitate the reaction rate.

11.2 Reverse osmosis systems

11.2.1 Reverse osmosis–nanofiltration cascade

In the following example, RO and nanofiltration (NF) modules are combined in the RO–NF cascade for the removal of sodium chloride. When both RO and NF are in the complete mixing mode (Fig. 11.2A), the flow rate and sodium chloride concentration at the first- and second-stage permeate can be obtained as follows. (It should be noted that concentration polarization is ignored in the following model development.)

Using the symbols given in Fig. 11.4.

Step 1: c_R^1 is assumed.

Step 2: c_p^1 is obtained.

From the water and salt transport through the membrane in the first module,

$$u_p^1 = S^1 J_A^1 = S^1 A^1 \left(\Delta p^1 - b c_R^1 + b c_p^1 \right) \tag{11.1}$$

$$u_p^1 c_p^1 = S^1 J_B^1 = S^1 B^1 \left(c_R^1 - c_p^1 \right) \tag{11.2}$$

where S^1, A^1, B^1, J_A^1, and J_B^1 are membrane area (m^2), pure water permeation coefficient (m^3/m^2 s kPa), solute permeation constant (m/s), flux of water (m^3/m^2 s), and flux of solute (mol/m^2 s), respectively, all of the first stage. b is the osmotic pressure coefficient. Superscript 1 means the first stage.

Solving Eqs. (11.1) and (11.2) for c_p^1,

$$c_p^1 = \frac{-(B^1 - A^1 b c_R^1 + A^1 \Delta p^1) + \sqrt{(B^1 - A^1 b c_R^1 + A^1 \Delta p^1)^2 + 4A^1 b B^1 c_R^1}}{2A^1 b} \tag{11.3}$$

From the volume flow balance,

$$u_R^1 = u_f - u_p^1 \tag{11.4}$$

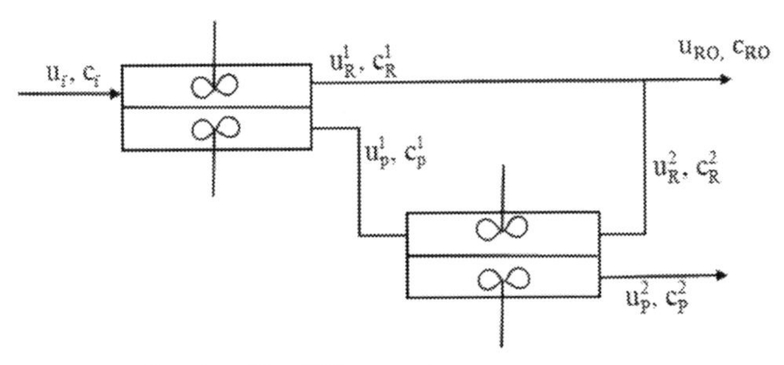

Figure 11.4 Reverse osmosis–nanofiltration (RO–NF) cascade.

From the salt flow balance,

$$c_R^1 = \frac{\left(u_f c_f - u_p^1 c_p^1\right)}{u_R^1} \tag{11.5}$$

Check if c_R^1 is the same as c_R^1 assumed in step 1. If not, go back to step 1. If they are the same continue to step 3.

Step 3: c_R^2 is assumed.

Step 4: c_p^2 is obtained.

From the water and salt transport through the membrane in the second module,

$$u_p^2 = S^2 J_A^2 = S^2 A^2 \left(\Delta p^2 - b c_R^2 + b c_p^2\right) \tag{11.6}$$

$$u_p^2 c_p^2 = S^2 J_B^2 = S^2 B^2 \left(c_R^2 - c_p^2\right) \tag{11.7}$$

where S^2, A^2, B^2, J_A^1, and J_B^2 are membrane area (m^2), pure water permeation coefficient (m^3/m^2 s kPa), solute permeation constant (m/s), flux of water (m^3/m^2 s), and flux of solute (mol/m^2 s), respectively, all of the second stage (with superscript 2).

Solving Eqs. (11.6) and (11.7) for c_p^2,

$$c_p^2 = \frac{-\left(B^2 - A^2 b c_R^2 + A^2 \Delta p^2\right) + \sqrt{\left(B^2 - A^2 b c_R^2 + A^2 \Delta p^2\right)^2 + 4A^2 b B^2 c_R^2}}{2A^2 b} \tag{11.8}$$

From the volume flow balance,

$$u_R^2 = u_p^1 - u_p^2 \tag{11.9}$$

From the salt balance,

$$c_R^2 = \frac{\left(u_p^1 c_p^1 - u_p^2 c_p^2\right)}{u_R^2} \tag{11.10}$$

Check if c_R^2 is the same as c_R^2 assumed in step 3. If not, go back to step 3. If they are the same continue to step 5.

Step 5: Calculate u_{RO} and c_{RO}.

$$u_{RO} = u_R^1 + u_R^2 \tag{11.11}$$

$$c_{RO} = \frac{\left(u_R^1 c_R^1 + u_R^2 c_R^2\right)}{u_{RO}} \tag{11.12}$$

The simulation algorithm is shown in Scheme 11.1.

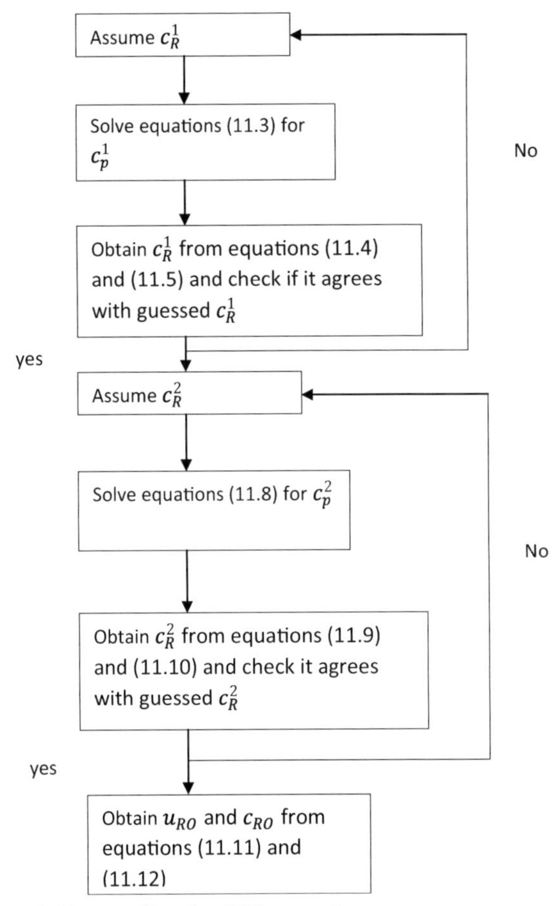

Scheme 11.1 Reverse osmosis (RO)–nanofiltration (NF) cascade.

Problem 11.1:

Calculate the flow rate and concentration of permeate at the first and the second stages of the RO–NF cascade, assuming complete mixing of both feed and permeate. Calculate also the flow rate and concentration of the retentate coming out of the system. Use the following design parameters.

Pure water permeation coefficient of RO membrane: $A^1 = 3.604 \times 10^{-10}\ \mathrm{m^3/m^2s\ kPa}$

Solute permeation constant of RO membrane: $B^1 = 0.075 \times 10^{-7}\ \mathrm{m/s}$

Pure water permeation coefficient of NF membrane: $A^2 = 6.087 \times 10^{-9}\ \mathrm{m^3/m^2s\ kPa}$

Solute permeation constant of NF membrane: $B^2 = 0.7193 \times 10^{-7}\ \mathrm{m/s}$

Flow rate of the first-stage feed: $u_f = 3 \times 10^{-6}\ \mathrm{m^3/s}$

Sodium chloride concentration of the first-stage feed: $c_f = 0.1\mathrm{kmol/m^3}$

Pressure applied at the first stage: $\Delta p^1 = 2758$ kPa(gauge)

Pressure applied at the second stage: $\Delta p^2 = 100$ kPa(gauge)

Osmotic pressure coefficient: $b = 4.64 \times 10^3\ \mathrm{kPa/(kmol/m^3)}$

Membrane area: $S^1$and $S^2 = 1\mathrm{m}^2$

Answer:

Step 1: c_R^1 is assumed to be 0.2 $\mathrm{kmol/m^3}$.

Step 2: From Eq. (11.3)

$$\alpha = B^1 - A^1 b c_R^1 + A^1 \Delta p^1 = 0.075 \times 10^{-7} - 3.604 \times 10^{-10} \times 4.64 \times 10^3 \times 0.2 + 3.604 \times 10^{-10} \times 2758 = 6.670 \times 10^{-7}$$

$$c_p^1 = \frac{-\alpha + \sqrt{\alpha^2 + 4 \times 3.604 \times 10^{-10} \times 4.64 \times 10^3 \times 0.075 \times 10^{-7} \times 0.2}}{2 \times 3.604 \times 10^{-10} \times 4.64 \times 10^3} = 2.24 \times 10^{-3}$$

From Eq. (11.1)

$$u_p^1 = 1 \times 3.604 \times 10^{-10} \times (2758 - 4.64 \times 10^3 \times 0.2 + 4.64 \times 10^3 \times 2.24 \times 10^{-3}) = 6.63 \times 10^{-7}$$

From Eq. (11.4)

$$u_R^1 = 3 \times 10^{-6} - 6.63 \times 10^{-7} = 2.37 \times 10^{-6}$$

From Eq. (11.5)

$$c_R^1 = \frac{(3 \times 10^{-6} \times 0.1 - 6.63 \times 10^{-7} \times 2.24 \times 10^{-3})}{2.37} \times 10^{-6} = 0.126 \neq 0.2$$

Therefore we go back to step 1.

After repeating steps 1 and 2, it is found that $c_R^1 = 0.134$ is the right answer. Corresponding to this c_R^1, u_R^1, u_p^1, and c_p^1 are $2.23 \times 10^{-6}\mathrm{m^3/s}, 7.72 \times 10^{-7}\mathrm{m^3/s}$, and $1.29 \times 10^{-3}\mathrm{kmol/m^3}$, respectively.

Steps 3 and 4:

By iteration, it is found that

$$u_R^2 = 2.44 \times 10^{-7}$$

$$c_R^2 = 3.24 \times 10^{-3}$$

Correspondingly,

$$u_p^2 = 5.28 \times 10^{-7}$$

$$c_p^2 = 3.88 \times 10^{-4}$$

Step 5:
From Eq. (11.11)

$$u_{RO} = 2.23 \times 10^{-6} + 2.44 \times 10^{-7} = 2.47 \times 10^{-6}$$

From Eq. (11.12)

$$c_{RO} = \frac{(2.23 \times 10^{-6} \times 0.134 + 2.44 \times 10^{-7} \times 3.24 \times 10^{-3})}{2.47} \times 10^{6} = 0.121$$

The answers are:

The first-stage permeate flow rate and concentration are $7.72 \times 10^{-7} m^3/s$ and 1.29×10^{-3} kmol/m^3.

The second-stage permeate flow rate and concentration are 5.28×10^{-7} m^3/s and 3.88×10^{-4} kmol/m^3.

The retentate flow rate and concentration are 2.47×10^{-6} m^3/s and 0.121 kmol/m^3.

11.2.2 Reverse osmosis recycle

In the following example, a fraction of the RO permeate is recycled to the feed inlet. When RO is in complete mixing mode, the flow rate of the permeate and retentate and their concentration can be calculated as follows. Again, concentration polarization is ignored.

Using the symbols given in Fig. 11.5.

Step 1: Assume c_R.

Step 2: c_p is obtained.

$$u_p = SJ_A = SA(\Delta p - bc_R + bc_p) \quad (11.13)$$

$$u_p c_p = SJ_B = SB(c_R - c_p) \quad (11.14)$$

Solving Eqs. (11.13) and (11.14) for c_p,

$$c_p = \frac{-(B - Abc_R + A\Delta p) + \sqrt{(B - Abc_R + A\Delta p)^2 + 4AbBc_R}}{2Ab} \quad (11.15)$$

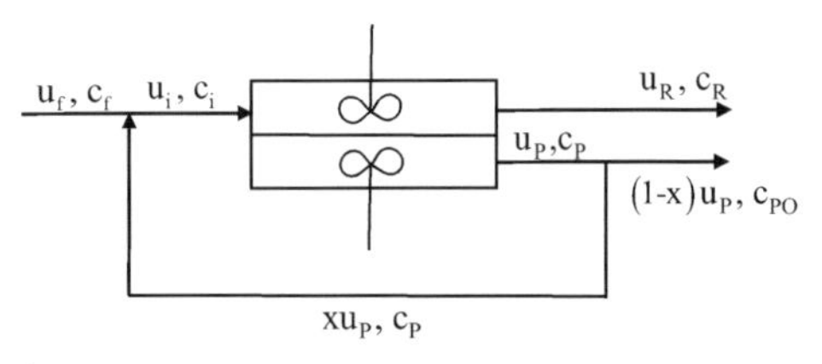

Figure 11.5 RO permeate recycled.

From Eq. (11.14)

$$u_p = \frac{SB(c_R - c_p)}{c_p} \tag{11.16}$$

$$u_R = u_f - (1-x)u_p \tag{11.17}$$

$$c_R = \frac{(u_f c_f - (1-x)u_p c_p)}{u_R} \tag{11.18}$$

Check if c_R is the same as c_R assumed in step 1. If not, go back to step 1. If they are the same continue to step 3.

Step 3: Calculate u_i, c_i, u_{po} and c_{po}.

$$u_i = u_p + u_R \tag{11.19}$$

$$c_i = \frac{(u_p c_p + u_R c_R)}{u_i} \tag{11.20}$$

$$u_{po} = (1-x)u_p \tag{11.21}$$

$$c_{po} = c_p \tag{11.22}$$

Simulation can be done using Scheme 11.2.

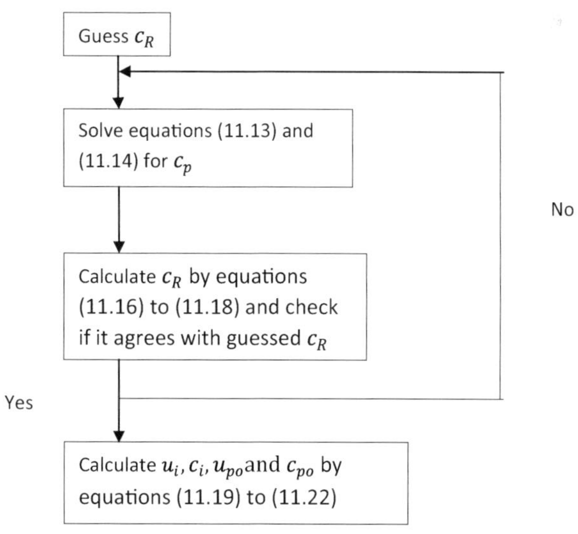

Scheme 11.2 Reverse osmosis recycle.

Problem 11.2:

Calculate the flow rate and concentration of the permeate coming out of the system and those of retentate, when a part of the permeate is recycled. Complete mixing of feed and permeate in the module is assumed. Use the following design parameters.

Pure water permeation coefficient of RO membrane: $A = 3.604 \times 10^{-10}$ m^3/m^{2}s kPa

Solute permeation constant of RO membrane: $B = 0.075 \times 10^{-7}$ m/s

Feed flow rate: $u_f = 5 \times 10^{-7}$ m^3/s

Sodium chloride concentration of the feed: $c_f = 0.1$ kmol/m^3

Pressure: $\Delta p = 2758$ kPa

Osmotic pressure coefficient: $b = 4.64 \times 10^3$ kPa/(kg/m^3)

Membrane area: $S = 1$ m^2

90% of permeate recycled to the feed: $x = 0.9$

Answer:

Step 1: c_R is assumed to be 0.2.

Step 2: From Eq. (11.15)

$$\alpha = B - Abc_R + A\Delta p = 0.075 \times 10^{-7} - 3.604 \times 10^{-10} \times 4.64 \times 10^3 \times 0.2 + 3.604 \times 10^{-10} \times 2758 = 6.670 \times 10^{-7}$$

$$c_p = \frac{-\alpha + \sqrt{\alpha^2 + 4 \times 3.604 \times 10^{-10} \times 4.64 \times 10^3 \times 0.075 \times 10^{-7} \times 0.2}}{2 \times 3.604 \times 10^{-10} \times 4.64 \times 10^3} = 2.24 \times 10^{-3}$$

$$u_p = \frac{1 \times 0.075 \times 10^{-7} \times (0.2 - 0.00224)}{0.00224} = 6.62 \times 10^{-7}$$

From Eqs. (11.17) and (11.18)

$$u_R = 5 \times 10^{-7} - (1 - 0.9) \times 6.62 \times 10^{-7} = 4.34 \times 10^{-7}$$

$$c_R = \frac{(5 \times 10^{-7} \times 0.1 - (1 - 0.9) \times 6.62 \times 10^{-7} \times 2.24 \times 10^{-3})}{4.34} \times 10^7 = 0.115 \neq 0.2$$

Therefore we go back to step 1. After repeating steps 1 and 2 we find that $c_R = 0.119$ is the right answer. Correspondingly, u_R, u_p, and c_p are 4.20×10^{-7}, 7.97×10^{-7}, and 0.00111, respectively.

Step 3: u_i and c_i and u_{po} and c_{po} are calculated from Eqs. (11.19)–(11.22).

$$u_i = 7.97 \times 10^{-7} + 4.20 \times 10^{-7} = 1.22 \times 10^{-6}$$

$$c_i = \frac{(7.97 \times 10^{-7} \times 0.00111 + 4.20 \times 10^{-7} \times 0.119)}{1.22 \times 10^{-6}} = 0.0417$$

$$u_{po} = (1 - 0.9) \times 7.97 \times 10^{-7} = 7.97 \times 10^{-8}$$

$$c_{po} = c_p = 0.00111$$

Problem 11.3:

Obtain u_{po} and c_{po} for different x.

Answer:

Fig. 11.6 shows u_{po} and c_{po} for different values of x.

11.3 Gas separator systems

11.3.1 Gas separator cascade

In the following example, two hollow fiber modules are connected in series to separate the gas mixtures A and B. In the module feed flows in the shell side and permeate flows co-currently in

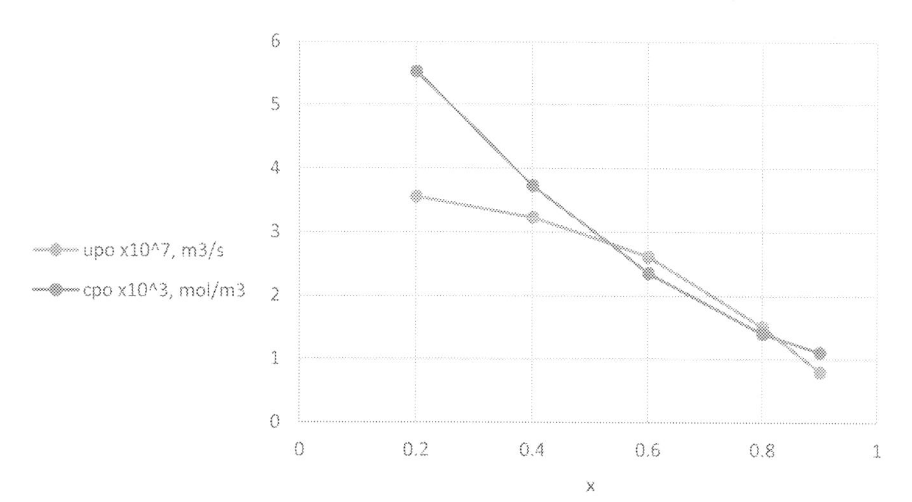

Figure 11.6 u_{po} and c_{po} versus x.

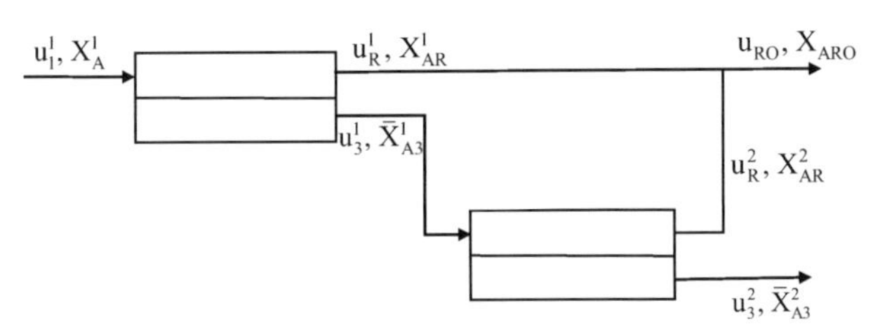

Figure 11.7 Gas separator connected in cascade.

the lumen side. The feed composition keeps changing from the hollow fiber entrance to the exit.

The simulation is the repeat of that given in Problem 10.3 in Chapter 10, Membrane Module, for two stages with superscripts indicating the stages.

Symbols and simulation algorithm are shown in Fig. 11.7 and Scheme 11.3, respectively.

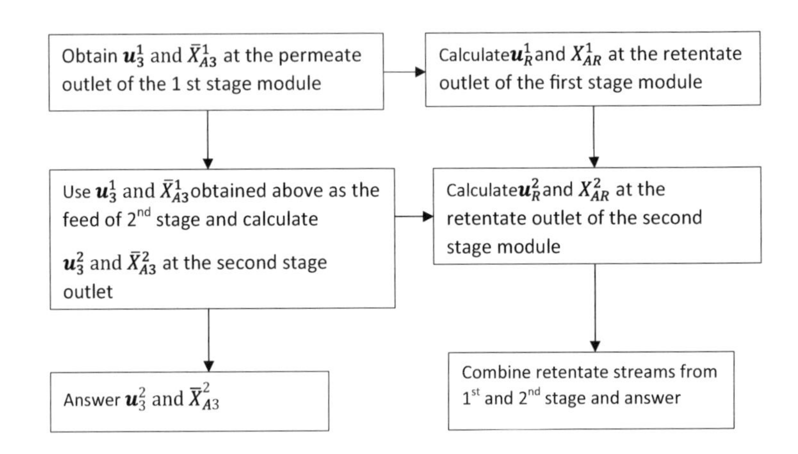

Scheme 11.3 Gas separator cascade.

Problem 11.4:

Calculate the molar flux $\boldsymbol{u}_3^2$ and mole fraction $\overline{X}_{A3}^2$ of the permeate and the molar flux $\boldsymbol{u}_{RO}$ and mole fraction X_{ARO} at the end of the hollow fiber module cascade, using the following parameters. The temperature is 35°C.

Molar feed gas velocity at the sealed end of hollow fiber: $\boldsymbol{u}_1(0) = 200\ \text{mol/m}^2\text{s}$

Feed CO_2 mole fraction at the sealed end of hollow fiber: $X_{A1}(0) = 0.2$

Feed gas pressure in the first stage: $p_1^1(0) = 10$ atmospheric pressure 1.01325×10^6 Pa (absolute)

Feed pressure in the second stage: $p_1^2(0) = 2$ atmospheric pressure $= 2.0265 \times 10^6$ Pa (absolute)

The same hollow fiber module is used in stages 1 and 2 with,
CO_2 permeance: $\frac{P_A}{\delta} = 26.3$ GPU $= 89.2 \times 10^{-10}$ mol/m^{2}s Pa
N_2 permeance: $\frac{P_B}{\delta} = 0.93$ GPU $= 3.15 \times 10^{-10}$ mol/m^{2}s Pa
Effective diameter of hollow fiber: $d_e = 1.05 \times 10^{-4}$ m
Inner diameter of hollow fiber: $d_i = 1.00 \times 10^{-4}$ m
Hollow fiber length: $L = 0.6$ m
Furthermore, the total cross-sectional area of the lumen side is the same as that of the shell side.
Viscosity of CO_2/N_2 gas mixture: $\mu = 1.663 \times 10^{-5}$ Pa s

Answer:

The conditions for the first stage are the same as those for Problem 10.3 in Chapter 10. Therefore the permeate comes out from stage 1 with

$$\boldsymbol{u}_3^1(0.6) = 25.8 \text{ mol/m}^2\text{s}$$

$$\overline{X}_{A3}^1(0.6) = 0.737$$

$$p_3^1(0.6) = 101.3 \text{ kPa}$$

And the retentate of stage 1 with

$$\boldsymbol{u}_R^1(0.6) = 200 - 25.8 = 174.2 \text{ mol/m}^2\text{s}$$

$$X_{\text{AR}}^1(0.6) = \frac{200 \times 0.2 - 25.8 \times 0.737}{174.2} = 0.121$$

The exit gas mixture from the first stage is pressurized to 2 bar and enters into the shell side of the second stage as feed. Since the area ratio of the shell side to lumen side is 1.0, in the second stage,

$$\boldsymbol{u}_1^2(0) = 25.8 \text{ mol/m}^2 \text{ s}$$

$$X_{A1}^2(0) = 0.737$$

Repeating the calculation for the lumen side of the hollow fiber,

$$\boldsymbol{u}_3^2(0.6) = 9.61 \text{ mol/m}^2\text{s}$$

$$\overline{X}_{A3}^2(0.6) = 0.949$$

$$p_3^2(0.6) = 101.3 \text{ kPa}$$

and

$$u_R^2(0.6) = 25.8 - 9.61 = 16.2 \text{ mol/m}^2\text{s}$$

$$X_{AR}^2(0.6) = \frac{25.8 \times 0.737 - 9.61 \times 0.949}{16.2} = 0.611$$

$$u_{RO} = 174.2 + 16.2 = 190.4 \text{ mol/m}^2 \text{ s}$$

$$X_{ARO} = (174.2 \times 0.121 + 16.2 \times 0.611)/190.4 = 0.163$$

11.3.2 Gas separator recycle

In the following example, a part of the permeate coming from the hollow fiber lumen side is recycled to the feed side of the hollow fiber. The feed and permeate flow co-currently. The permeate flow in the lumen side is assumed to be piston flow. The flow rate and concentration of the permeate and those of the retentate can be calculated as follows. (Note that the cross-sectional area of the shell side S_1 and that of the shell side S_3 are assumed to be equal.)

Using the symbols given in Fig. 11.8,

Step 1: $\boldsymbol{u}_i$ and X_{Ai} are assumed as the initial guess.

Step 2: $p_3(0)$ is assumed. Follow step 2 of Problem 10.3 in Chapter 10 until $p_3(L)$ becomes the hollow fiber exit pressure that is equal to atmospheric pressure (1.013×10^5 Pa).

$\boldsymbol{u}_3(L)$ and $\overline{X}_{A3}(L)$ that correspond to the initial guess are also obtained.

Then,

$$\boldsymbol{u}_i = \boldsymbol{u}_f + x\boldsymbol{u}_3(L) \tag{11.23}$$

Since $S_1 = S_3$ and

$$X_{Ai} = \frac{\left(\boldsymbol{u}_f X_{Af} + x\boldsymbol{u}_3(L)\overline{X}_{A3}(L)\right)}{\boldsymbol{u}_i} \tag{11.24}$$

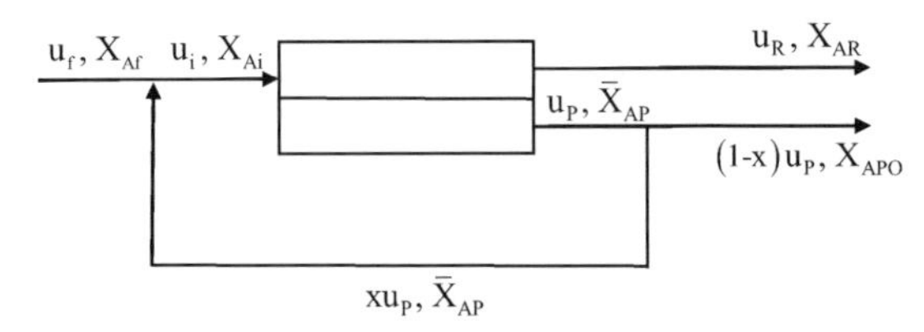

Figure 11.8 Gas separator recycle.

Check if these $\boldsymbol{u}_i$ and X_{Ai} agree with the guessed $\boldsymbol{u}_i$ and X_{Ai}. If they agree go to step 3. Otherwise, go back to step 1.

Step 3:

Obtain $\boldsymbol{u}_R$ and X_{AR}

$$\boldsymbol{u}_R = \boldsymbol{u}_f - (1-x)\boldsymbol{u}_3(L) \tag{11.25}$$

$$X_{AR} = \frac{\left(\boldsymbol{u}_f X_{Af} - (1-x)\boldsymbol{u}_3(L)\overline{X}_{A3}(L)\right)}{(1-x)\boldsymbol{u}_3(L)} \tag{11.26}$$

$$\boldsymbol{u}_{po} = (1-x)\boldsymbol{u}_3(L) \tag{11.27}$$

$$X_{Apo} = \overline{X}_{A3}(L) \tag{11.28}$$

The simulation algorithm is given in Scheme 11.4.

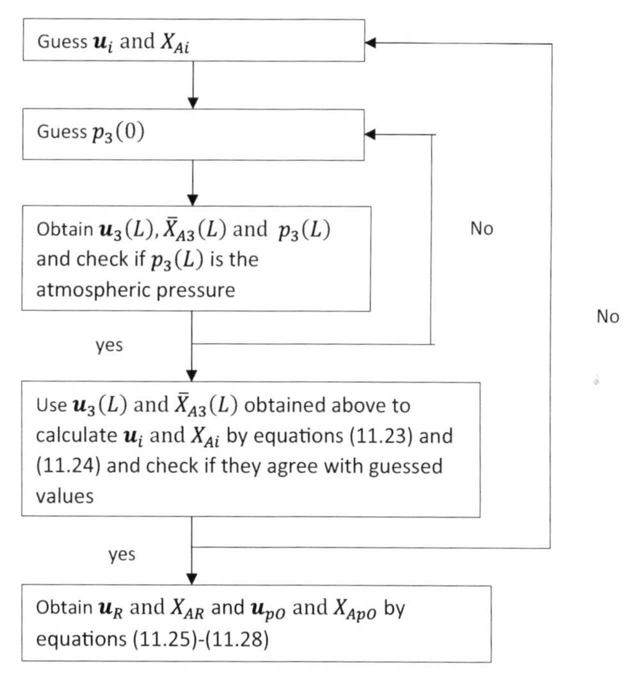

Scheme 11.4 Gas separator recycle.

Problem 11.5:

Calculate $\boldsymbol{u}_R$, X_{AR}, $\boldsymbol{u}_{PO}$, and X_{APO} for $x = 0.5$, using the following parameters.

Molar feed gas velocity before being mixed with the recycle gas: $\boldsymbol{u}_f = 200\ \text{mol/m}^2\text{s}$

Feed CO_2 mole fraction before being mixed with the recycle gas: $X_{Af}(0) = 0.2$

Feed gas pressure: $p_1(0) = 10$ atmospheric pressure 1.01325×10^6 Pa (absolute)

CO_2 permeance: $P_A/\delta = 26.3$ GPU $= 89.2 \times 10^{-10}$ mol/m²sPa

N_2 permeance: $P_B/\delta = 0.93$ GPU $= 3.15 \times 10^{-10}$ mol/m²sPa

Effective diameter of hollow fiber: $d_e = 1.05 \times 10^{-4}$ m

Inner diameter of hollow fiber: $d_i = 1.00 \times 10^{-4}$ m

Hollow fiber length: $L = 0.6$ m

Viscosity of CO_2/N_2 gas mixture: $\mu = 1.663 \times 10^{-5}$ Pa s

$$S_1 = S_3$$

These parameters are the same as those of Problem 10.3 in Chapter 10.

Answer:

Step 1: $\boldsymbol{u}_i$ and X_{Ai} are assumed to be 216 and 0.244, respectively.

Step 2:

Step 2.1: $p_3(0)$ is assumed to be 1.12×10^5 Pa.

$J_A(0), J_B(0), X_{A3}(0)$ are obtained by solving Eqs. (10.63), (10.64), (10.69), and (10.70), where p_i and X_{Ai} are substituted for p_1 and X_{A1}.

Using the given numerical values and solving,

$$\alpha = 0.001768$$

$$X_{A3}(0) = 0.852$$

From Eq. (10.63)

$$J_A = 89.2 \times 10^{-10} \times \left(1.01325 \times 10^6 \times 0.244 - 1.12 \times 10^5 \times 0.852\right)$$
$$= 1.35 \times 10^{-3}$$

From Eq. (10.64)

$$J_B = 3.15 \times 10^{-10} \times \left(1.01325 \times 10^6 \times 0.756 - 1.12 \times 10^5 \times 0.148\right)$$
$$= 2.36 \times 10^{-4}$$

Thus,

$$\overline{X}_{A3}(0) = X_{A3}(0) = 0.852$$

Step 2.2: The derivative $(d\boldsymbol{u}_3/dx)_{x=0}$ is obtained.

From Eq. (10.53)

$$\frac{d\boldsymbol{u}_3(0)}{dx} = \frac{4 \times 1.05 \times 10^{-4}}{\left(1.00 \times 10^{-4}\right)^2}\left(1.35 \times 10^{-3} + 2.36 \times 10^{-4}\right) = 66.8$$

From Eq. (10.54)

$$\frac{d\overline{X}_{A3}(0)}{dx} = 0$$

From Eq. (10.56)

$$\frac{dp_3(0)}{dx} = 0$$

Step 2.3: $\boldsymbol{u}_3(\Delta x), \overline{X}_{A3}(\Delta x)$, and $p_3(\Delta x)$ are obtained for a pre-set size of the increment $\Delta x = 0.1$.

($\Delta x = 0.1$ may be too large. But this was used only to explain how the problem is solved.)

$$\boldsymbol{u}_3(0.1) = 0 + 66.8 \times 0.1 = 6.68$$

$$\overline{X}_{A3}(0.1) = 0.852 + 0 = 0.852$$

$$p_3(0.1) = 1.12 \times 10^5 + 0 = 1.12 \times 10^5$$

Step 3: $\boldsymbol{u}_3(0.2)$, $\overline{X}_{A3}(0.2)$, and $p_3(0.2)$ are obtained.

Step 3.1: $J_A(0.1), J_B(0.1), X_{A3}(0.1)$ are obtained by solving Eqs. (10.58), (10.63), (10.64), (10.69), and (10.70)

$$X_{A1}(0.1) = \frac{\left(\boldsymbol{u}_1(0)X_{A1}(0) - \boldsymbol{u}_3(0.1)\overline{X}_{A3}(0.1)\right)}{(\boldsymbol{u}_1(0) - \boldsymbol{u}_3(0.1))} = \frac{(216 \times 0.244 - 6.68 \times 0.852)}{216 - 6.68} = 0.225$$

$$\alpha = \sqrt{(P_A p_1 X_{A1} + P_B p_1 X_{B1} + P_A p_3 - P_B p_3)^2 + 4 \times (P_B p_3 - P_A p_3) \times P_A p_1 X_{A1}}$$

$$\alpha = 0.001637$$

$$X_{A3}(0.1) = 0.832$$

$$J_A(0.1) = 1.20 \times 10^{-3}$$

$$J_B(0.1) = 2.42 \times 10^{-4}$$

From Eq. (10.53)

$$\frac{d\boldsymbol{u}_3(0.1)}{dx} = \frac{4 \times 1.05 \times 10^{-4}}{(1.00 \times 10^{-4})^2}\left(1.20 \times 10^{-3} + 2.42 \times 10^{-4}\right) = 60.5$$

From Eq. (10.55)

$$\frac{d\overline{X}_{A3}(0.1)}{dx} = \frac{4 \times 1.05 \times 10^{-4}}{6.68 \times (1.00 \times 10^{-4})^2}\{1.20 \times 10^{-3} \times 0.148 - 2.42 \times 10^{-4} \times 0.852\}$$
$$= -0.175$$

From Eq. (10.56)

$$\frac{dp_3(x)}{dx} = \frac{64\mu RT}{2p_3 d_i^2}\boldsymbol{u}_3$$

$$\frac{dp_3(0.1)}{dx} = -\frac{64 \times 1.66 \times 10^{-5} \times 8.314 \times 308.2}{2 \times 1.12 \times 10^5 \times (1.00 \times 10^{-4})^2} \times 6.68 = -8130$$

$$\boldsymbol{u}_3(0.2) = 6.68 + 60.5 \times 0.1 = 12.73$$

$$\overline{X}_{A3}(0.2) = 0.852 - 0.175 \times 0.1 = 0.8345$$

$$p_3(0.2) = 1.12 \times 10^5 - 8130 \times 0.1 = 1.11 \times 10^5$$

$p_3(0.6) = 1.01 \times 10^5$, which agrees with the guess of $p_3(0.6)$.

The final $\boldsymbol{u}_3(0.6)$ and $\overline{X}_{A3}(0.6)$ obtained are, respectively, 32.2 and 0.797.

From Eqs. (11.23) and (11.24)

$$\boldsymbol{u}_i = 200 + 0.5 \times 32.2 = 216.1$$

$$X_{Ai} = \frac{(200 \times 0.2 + 0.5 \times 32.2 \times 0.797)}{216.1} = 0.244$$

Both agree with the guessed $\boldsymbol{u}_i$ and X_{Ai}.

Step 3: From Eqs. (11.25)–(11.28),

$$\boldsymbol{u}_R = 200 - (1 - 0.5) \times 32.2 = 183.9$$

$$X_{AR} = \frac{(200 \times 0.2 - (1 - 0.5) \times 32.2 \times 0.797)}{183.2} = 0.148$$

$$\boldsymbol{u}_{PO} = (1 - 0.5) \times 32.2 = 16.1$$

$$X_{APO} = 0.797$$

11.4 Forward osmosis–reverse osmosis hybrid system

Forward osmosis (FO)–RO hybrid system is used for simultaneous wastewater treatment and seawater desalination.

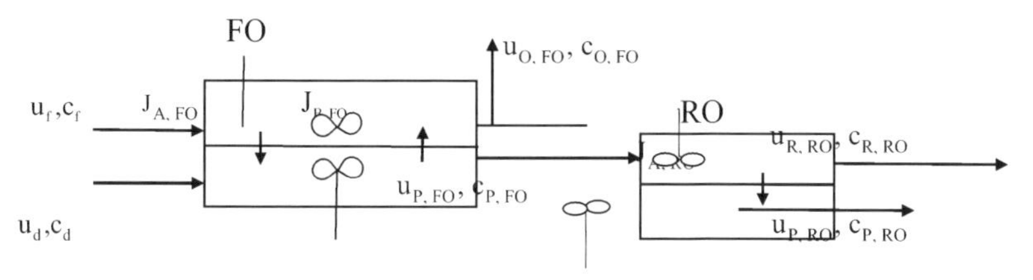

Figure 11.9 Forward osmosis–reverse osmosis (FO–RO) hybrid system.

Dilution of seawater by FO alleviates the high-pressure requirement of RO. Hence, wastewater and seawater are provided into FO as the feed and draw solution (DS), respectively, to dilute the seawater (Hancock et al., 2012). Then, the diluted seawater is subjected to RO to produce clean potable water, as shown schematically in Fig. 11.9. The system has been already used in desalination of water from the Red Sea (Yaeli, 1992). Experiments have also been made for an FO–RO system where three SWROs were connected in series and combined with FO (Cath et al., 2010). One drawback of the FO–RO hybrid system is the rather low FO permeate flux. To increase water recovery and to reduce the amount of brine discharge, Altaee and Hilal proposed the NF-FO-BWRO process (Altaee and Hilal, 2015).

In the following example, a model simulation is made where FO and RO are connected in series. Both FO and RO are run in complete-stirred tank mode with membranes that have the same pure water permeation coefficient and solute permeation constant. FO is run in the AL-facing-DS mode. The internal concentration polarization in FO is considered while external concentration polarizations in both FO and RO are ignored.

The concentration and flow rate of the second RO stage permeate ($c_{p,RO}$ and $u_{p,RO}$) and those of the second RO stage retentate ($u_{R,RO}$ and $c_{R,RO}$) can be obtained as follows.

Step 1: Assume $J_{A,FO}$

Step 2: Calculate $u_{p,FO}$ and $c_{p,FO}$

Eqs. (3.54) and (3.55) of Chapter 3, Reverse Osmosis, Forward Osmosis, and Pressure-Retarded Osmosis, are

$$J_A = A\left(\pi_{DS} - \pi_{\frac{AL}{SL}}\right)(3.54)$$

$$J_B = B\left(c_{DS} - c_{\frac{AL}{SL}}\right)(3.55)$$

Assuming $\pi = bc$

$$J_{B,FO} = \frac{B_{FO}}{A_{FO}b} J_{A,FO} \quad (11.29)$$

Further,

$$u_{O,FO} = u_f - SJ_{A,FO} \quad (11.30)$$

$$c_{O,FO} = \frac{u_f c_f + SJ_{B,FO}}{u_{O,FO}} \quad (11.31)$$

$$u_{p,FO} = u_f + u_d - u_{O,FO} \quad (11.32)$$

$$c_{p,FO} = \frac{(u_f c_f + u_d c_d - u_{O,FO} c_{O,FO})}{u_{p,FO}} \quad (11.33)$$

Eq. (3.70) of Chapter 3 is

$$J_A = k_m ln\left(\frac{A\pi_{DS} - J_A + B}{A\pi_{FS} + B}\right)$$

for the AL-facing-DS mode operation.

Hence, using the symbols given in Fig. 11.9 and again assuming $\pi = bc$,

$$J_{A,FO} = k_m ln\left(\frac{A_{FO} b c_{p,FO} - J_{A,FO} + B_{FO}}{A_{FO} b c_{O,FO} + B_{FO}}\right) \quad (11.34)$$

Calculate $J_{A,FO}$ by Eq. (11.34) and check if it agrees with the guessed $J_{A,FO}$ value.

If they agree, go to step 3, otherwise go back to step 1.

Step 3: Assume $c_{R,RO}$.

Step 4: Obtain $c_{p,RO}$ by

$$c_{p,RO} = \frac{-(B_{RO} - A_{RO} b c_{R,RO} + A_{RO}\Delta p) + \sqrt{(B_{RO} - A_{RO} b c_{R,RO} + A_{RO}\Delta p)^2 + 4A_{RO} b B_{RO} c_{R,RO}}}{2A_{RO} b} \quad (11.35)$$

Then,

$$u_{p,RO} = SA_{RO}(\Delta p - b c_{R,RO} + b c_{p,RO}) \quad (11.36)$$

$$u_{R,RO} = u_{p,FO} - u_{p,RO} \quad (11.37)$$

$$c_{R,RO} = \frac{(u_{p,FO} c_{p,FO} - u_{p,RO} c_{p,RO})}{u_{R,RO}} \quad (11.38)$$

Check if $c_{R,\mathrm{RO}}$ so obtained agrees with the guessed $c_{R,\mathrm{RO}}$. If so, go to step 5. Otherwise, go back to step 3.

Step 5. Report $c_{p,RO}, u_{p,RO}, c_{R,RO}$, and $u_{R,RO}$.

The simulation algorithm is given in Scheme 11.5.

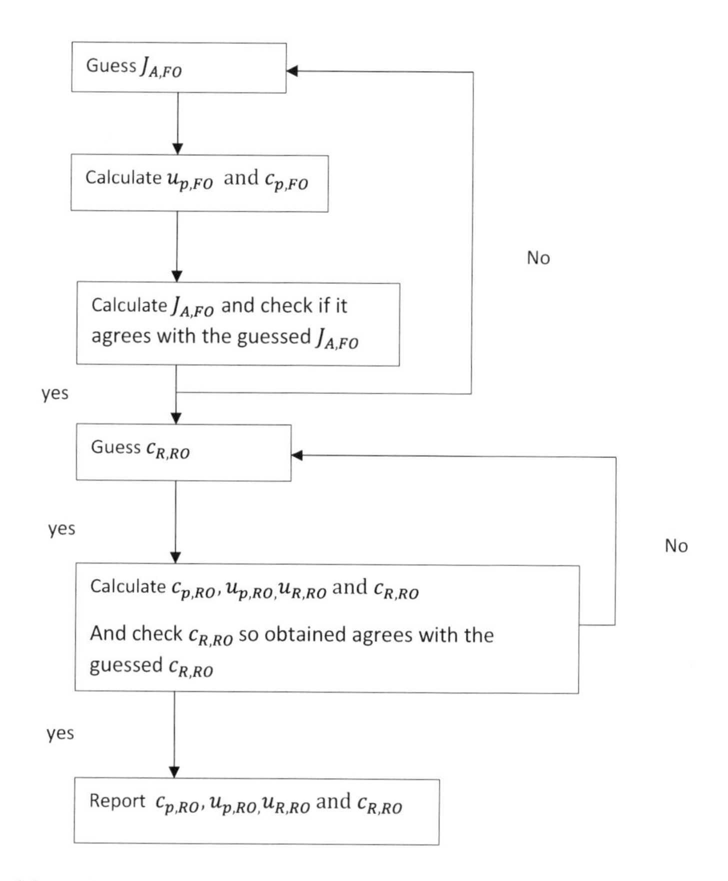

Scheme 11.5 FO–RO hybrid system.

Problem 11.6:

Obtain $c_{p,RO}$, $u_{p,RO}$, $c_{R,RO}$, and $u_{R,RO}$ of an FO–RO hybrid system with the following design parameters.

Pure water permeation coefficient of FO membrane: $A_{FO} = 6.083 \times 10^{-9} \text{m}^3/\text{m}^2\text{s kPa}$

Solute permeation constant of FO membrane: $B_{FO} = 71.93 \times 10^{-9}$ m/s

Pure water permeation coefficient of RO membrane: $A_{RO} = 6.083 \times 10^{-9}$ $\text{m}^3/\text{m}^2\text{s kPa}$

Solute permeation constant of RO membrane: $B_{RO} = 71.93 \times 10^{-9}$ m/s

Mass transfer coefficient: $k_m = 1.546 \times 10^{-6}$ m/s

Flow rate of the feed solution to FO: $u_f = 8 \times 10^{-6}$ m^3/s

Flow rate of the DS to FO: $u_d = 1 \times 10^{-7}$ m^3/s

Sodium chloride concentration of the feed solution in FO: $c_f = 0.001$ kmol/m^3(0.058%)

Sodium chloride concentration of the DS in FO: $c_d = 0.6$ kmol/m^3(3.5%)

Pressure applied at RO: $\Delta p = 250$ kPa
Osmotic pressure coefficient: $b = 4.64 \times 10^3$ kPa/(kmol/m^3)
FO and RO membrane area: $S = 1$ m^2
FO operation in the AL-facing-DS mode.

Answer:

Step 1: $J_{A,FO}$ is assumed to be 1.135×10^{-6}
Step 2: From Eq. (11.29)

$$J_{B,FO} = \frac{71.93 \times 10^{-9}}{6.083 \times 10^{-9} \times 4.64 \times 10^3} \times 1.135 \times 10^{-6} = 2.89 \times 10^{-9}$$

Further, from Eq. (11.30)

$$u_{O,FO} = 8 \times 10^{-6} - 1.135 \times 10^{-6} = 6.865 \times 10^{-6}$$

From Eq. (11.31)

$$c_{O,FO} = \frac{8 \times 10^{-6} \times 0.001 + 2.89 \times 10^{-9}}{6.865 \times 10^{-6}} = 1.586 \times 10^{-3}$$

From Eq. (11.32)

$$u_{p,FO} = u_f + u_d - u_{O,FO} = 8 \times 10^{-6} + 1 \times 10^{-7} - 6.865 \times 10^{-6} = 1.235 \times 10^{-6}$$

From Eq. (11.33)

$$c_{p,FO} = \frac{u_f c_f + u_d c_d - u_{O,FO} c_{O,FO}}{u_{p,FO}} = \frac{8\times10^{-6}\times0.001+1\times10^{-7}\times0.6-6.865\times10^{-6}\times1.586\times10^{-3}}{1.235\times10^{-6}} = 4.62 \times 10^{-2}$$

From Eq. (11.34)

$$J_{A,FO} = 1.546 \times 10^{-6} \ln\left(\frac{6.083 \times 10^{-9} \times 4.64 \times 10^3 \times 4.62 \times 10^{-2} - 1.135 \times 10^{-6} + 71.93 \times 10^{-9}}{6.083 \times 10^{-9} \times 4.64 \times 10^3 \times 1.586 \times 10^{-3} + 71.93 \times 10^{-9}}\right) = 1.14 \times 10^{-6}$$

$J_{A,FO}$ obtained is equal to the guessed $J_{A,FO}$. Therefore, we go to step 3.

Step 3: Assume $c_{R,RO} = 0.05$
Step 4:

$$\alpha = B_{RO} - A_{RO} b c_{R,RO} + A_{RO} \Delta p = 71.93 \times 10^{-9} - 6.087 \times 10^{-9} \times 4.64 \times 10^3 \times 0.05 + 6.087 \times 10^{-9} \times 250 = 18.15 \times 10^{-8}$$

$$c_{p,RO} = \frac{-\alpha+\sqrt{\alpha^2+4\times6.087 \times 10^{-9}\times4.64\times10^3\times71.93\times10^{-9}\times0.05}}{2 \times 6.087 \times 10^{-9} \times 4.64 \times 10^3} = 8.52 \times 10^{-3}$$

From Eq. (11.36)

$$u_{p,RO} = 1 \times 6.087 \times 10^{-9} \times \left(250 - 4.64 \times 10^{3} \times 0.05 + 4.64 \times 10^{3} \times 8.52 \times 10^{-3}\right)$$

$$= 3.50 \times 10^{-7}$$

From Eq. (11.37)

$$u_{R,RO} = u_{p,FO} - u_{p,RO} = 1.24 \times 10^{-6} - 3.50 \times 10^{-7} = 0.89 \times 10^{-6}$$

From Eq. (11.38)

$$c_{R,RO} = \frac{\left(u_{p,FO}c_{p,FO} - u_{p,RO}c_{p,RO}\right)}{u_{R,RO}}$$

$$= \frac{\left(1.24 \times 10^{-6} \times 4.62 \times 10^{-2} - 3.50 \times 10^{-7} \times 8.52 \times 10^{-3}\right)}{0.89} \times 10^{-6}$$

$$= 6.12 \times 10^{-2} > 0.05$$

We go back to step 3 and assume $c_{R,RO} = 0.056$
$c_{R,RO}$ obtained from Eq. (11.38) is 0.056.
Therefore, we can go to step 5.
Step 5: $c_{p,RO}$, $u_{p,RO}$, and $u_{R,RO}$ that correspond to $c_{R,RO} = 0.056$ are as follows.

$$c_{p,RO} = 0.0117\ \mathrm{kg/m^3}$$

$$u_{p,RO} = 2.71 \times 10^{-7}\ \mathrm{m^3/s}$$

$$u_{R,RO} = 9.64 \times 10^{-7}\mathrm{m^3/s}$$

11.5 Membrane reactor

The membrane reactor has been extensively studied for applications in fermentation processes as a membrane bioreactor. Although various membrane separation processes can be used in the membrane bioreactor, ultrafiltration (UF) and microfiltration are used in most cases in combination with a fermenter. The fermentation broth from a fermenter is allowed to flow continuously through a cross-flow membrane filtration module, where the microorganisms are rejected by the membrane and recycled back to the fermenter. As a result, a high cell density can be achieved in the fermenter. On the other hand, the product, such as alcohol, that acts as a reaction inhibitor is removed from the fermentation broth, enhancing the fermentation productivity. 2–5 g/L h alcohol are produced by the conventional batch fermenter, while

100 g/L h can be achieved by the continuous system in which membrane filtration is incorporated. However, the separation of substrate and products is not possible since both pass through the membrane.

Another membrane process that can be combined with a fermenter is pervaporation. The fermentation broth is supplied as the feed to a pervaporation module and a vacuum is applied to the permeate side of the membrane. When an ethanol-selective membrane is used the product ethanol passes through the membrane as vapor which is later liquefied by compression. Microorganisms, substrate, and nutrients are all retained on the feed side. Thus, the product ethanol is purified and concentrated. When the feed ethanol concentration was 5.5 wt.%, a permeate ethanol concentration of 30.6% could be achieved (Mulder and Smolders, 1966). Concentrated ethanol could further be dehydrated by pervaporation, using a water-selective membrane for pervaporation.

In the following example, it is shown that a RO membrane can be used to separate ethanol from microorganisms, substrates, and nutrients.

11.5.1 Description of the membrane bioreactor

A bioreactor with yeast cells immobilized between UF and RO membranes, both made of cellulose acetate, has been proposed (Vasdevan et al., 1987; Jeong et al., 1989). An aqueous solution containing glucose substrate and nutrients is in contact with the UF membrane and pressure is applied on the solution. The substrate permeates the UF membrane freely, together with water, and arrives at the cell layer where a bioreaction takes place to produce ethanol. Then, the ethanol permeates through the RO membrane, while glucose is retained in the yeast cell layer (see Fig. 11.10A; Matsuura, 1994).

The advantages of such a bioreactor over the conventional bioreactor are as follows:

1. High cell concentration within a limited reactor space;
2. Forced convective flow of substrate to the cell layer;
3. Removal of product ethanol and CO_2 from the cell layer, preventing product inhibition.

The continuous-type bioreactor system is schematically illustrated in Fig. 11.10B (Matsuura, 1994). After passing through the bioreactor, the substrate solution is recycled back to the feed tank.

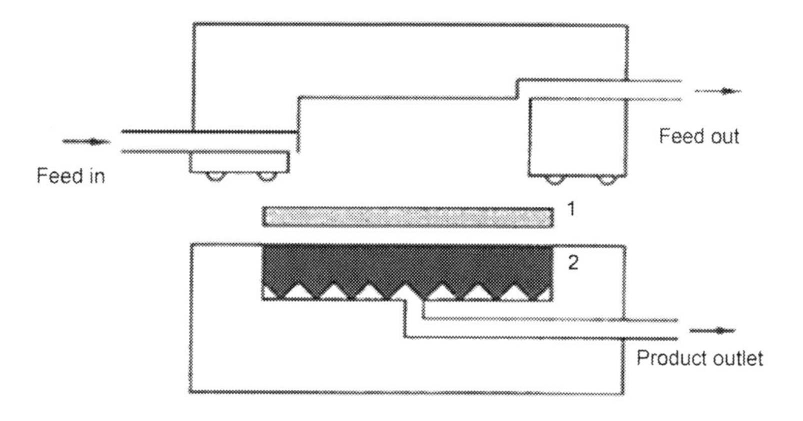

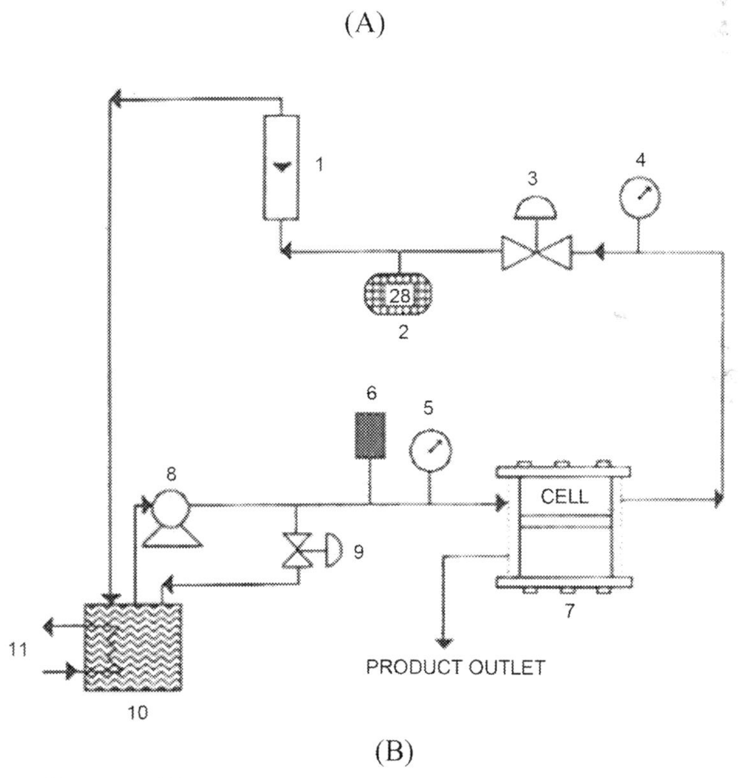

Figure 11.10 Membrane bioreactor: (A) permeation cell with sandwiched enzyme, (B) membrane bioreactor system; (1) flow meter, (2) thermometer, (3) back-pressure regulator, (4, 5) pressure gauge, (6) accumulator, (7) membrane bioreactor, (8) high-pressure pump, (9) valve, (10) feed tank, (11) heat exchanger (Matsuura, 1994).

11.5.2 Bioreactor modeling

For the system modeling a quasi-stationary approach is adopted with an assumption that the time required for glucose and ethanol concentration to reach the steady state is far shorter than the time required for the cell number increase. In order to facilitate the understanding of the model, the symbols defined in Fig. 11.11 are used, that is, *a*, *b*, *c*, and *d* are concentration polarization boundary layer, UF membrane, yeast cell layer, and RO

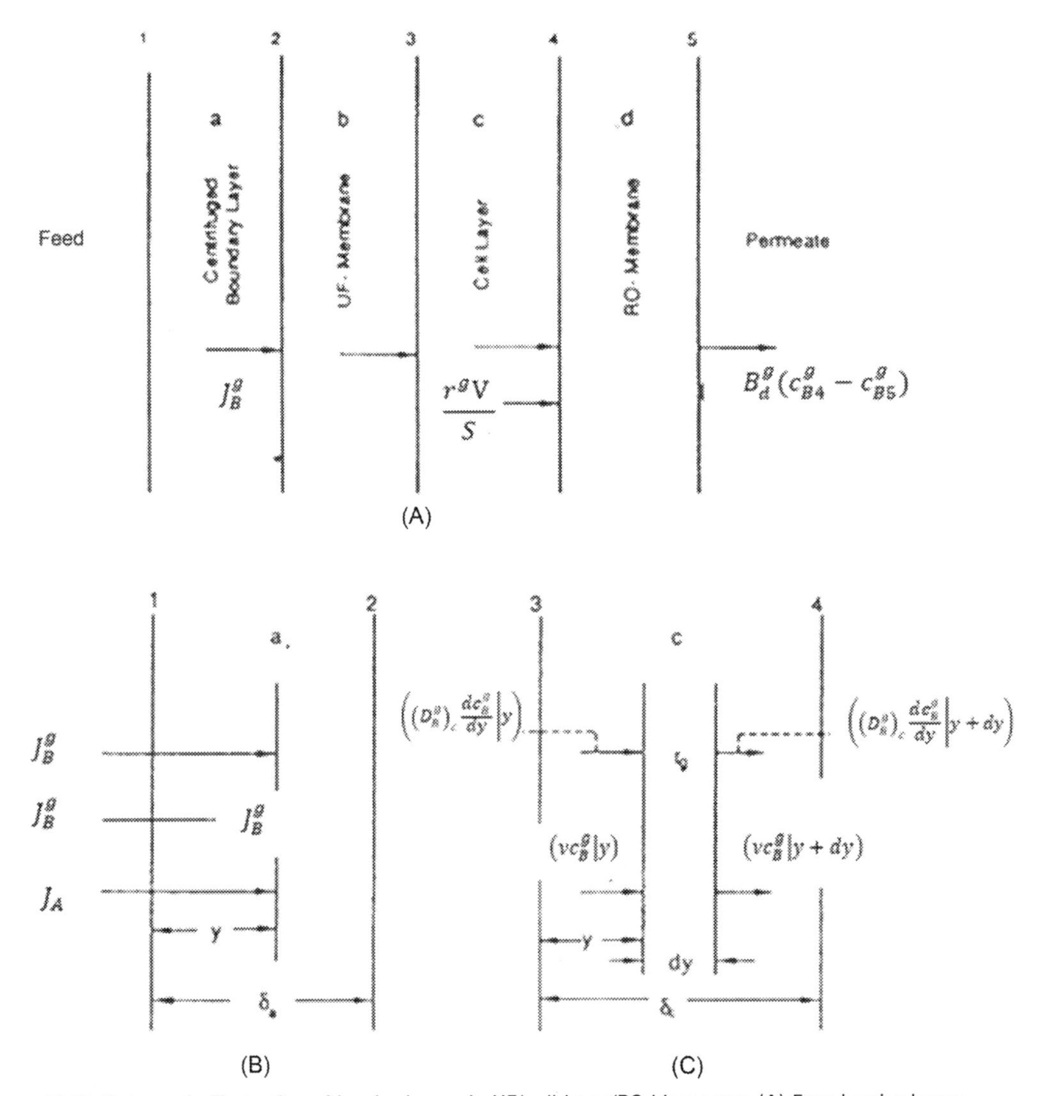

Figure 11.11 Schematic illustration of barrier layers in UF/cell layer/RO bioreactor. (A) Four barrier layers, (B) transport in barrier layer a, and (C) transport in barrier layer c (Matsuura, 1994).

membrane, respectively, which constitute four barrier layers; 1 and 5 are feed and permeate, respectively; 2, 3, and 4 are the interfaces between each barrier layer; and subscripts A and B are used for water and solute (either glucose or ethanol) as usual. All other symbols are defined at the end of this chapter.

The water fluxes through the barrier layer *b*, *c* and *d* are given, respectively, by

$$J_A = A_b(p_2 - p_3) \tag{11.39}$$

$$J_A = A_c(p_3 - p_4) \tag{11.40}$$

$$J_A = A_d(p_4 - p_5 - \pi_4 + \pi_5) \tag{11.41}$$

where J_A, A and p, respectively, are molar flux (mol/m^2 s), pure water permeation coefficient (mol/m^2 s Pa), and pressure (Pa).

As for the molar flux of glucose, it is the same through the barrier layers a and b, which is given by J_B^g (mol/m^2 s). In the cell layer (barrier layer c), it is consumed at the rate of r^g (mol/m^3 s, note r^g is negative). Suppose the yeast cell layer volume is V and the effective area of the membranes in which yeast cells are confined is S, the molar flux of glucose at the interface 4 becomes $J_B^g + r^g V/S$.

Approximating that the total molar flux is equal to the molar flux of water, which can be justified because the water flux is much higher than those of the solutes,

$$X_{B5}^g = \frac{J_B^g + \frac{r^g V}{S}}{J_A} \tag{11.42}$$

Furthermore, the permeation velocity (m/s)

$$v = \frac{\text{total molar flux}}{c} \approx \frac{J_A}{c} \tag{11.43}$$

where c is the total molar concentration (mol/m^3).

In the barrier layer *a* (concentration polarization boundary layer; see Fig. 11.11B), the molar flux of glucose at a distance *y* from the interface 1 is given by

$$J_B^g = -\left(D_B^g\right)_a c\frac{dX_B^g}{dy} + X_B^g\left(J_B^g + J_B^e + J_A\right) \tag{11.44}$$

where $\left(D_B^g\right)_a$ is the diffusivity of glucose (m^2/s) in the barrier layer *a* (i.e., in water).

Solving Eq. (11.44) with the boundary conditions

$$X_B^g = X_{B1}^g \quad \text{at } y = 0 \tag{11.45}$$

$$X_B^g = X_{B2}^g \quad \text{at y} = \delta_a \tag{11.46}$$

where δ_a is the thickness of the barrier layer a, we obtain

$$\ln \frac{X_{B2}^g - \left[\frac{J_B^g}{J_B^g + J_B^e + J_A}\right]}{X_{B1}^g - \left[\frac{J_B^g}{J_B^g + J_B^e + J_A}\right]} = \frac{J_B^g + J_B^e + J_A}{\left[\frac{(D_B^g)_a}{\delta_a}\right] c} \tag{11.47}$$

Approximating that total flux is nearly equal to water flux,

$$\ln \frac{X_{B2}^g - \left[\frac{J_B^g}{J_A}\right]}{X_{B1}^g - \left[\frac{J_B^g}{J_A}\right]} = \frac{J_A}{\left[\frac{(D_B^g)_a}{\delta_a}\right] c} \tag{11.48}$$

Combining Eqs. (11.42), (11.43), and (11.48)

$$\ln \frac{X_{B2}^g - X_{B5}^g + \left[\frac{r^g V/S}{J_A}\right]}{X_{B1}^g - X_{B5}^g + \left[\frac{r^g V/S}{J_A}\right]} = \frac{\nu}{k_a^g} \tag{11.49}$$

where k_a^g, the mass transfer coefficient (m/s) of glucose in the barrier layer a, is

$$k_a^g = \frac{(D_B^g)_a}{\delta_a} \tag{11.50}$$

Multiplying c on both the numerator and denominator of Eq. (11.49)

$$\ln \frac{c_{B2}^g - c_{B5}^g + \left[\frac{r^g V/S}{\nu}\right]}{c_{B1}^g - c_{B5}^g + \left[\frac{r^g V/S}{\nu}\right]} = \frac{\nu}{k_a^g} \tag{11.51}$$

Further

$$\frac{c_{B2}^g - c_{B5}^g + \left[\frac{r^g V/S}{\nu}\right]}{c_{B1}^g - c_{B5}^g + \left[\frac{r^g V/S}{\nu}\right]} = \exp\left(\frac{\nu}{k_a^g}\right) \tag{11.52}$$

A similar equation is applicable for the concentration polarization in the UF membrane (barrier layer b), therefore

$$\frac{c_{B3}^g - c_{B5}^g + \left[\frac{r^g V/S}{\nu}\right]}{c_{B1}^g - c_{B5}^g + \left[\frac{r^g V/S}{\nu}\right]} = \exp\left(\frac{\nu}{k_b^g}\right) \tag{11.53}$$

where k_b^g is the mass transfer coefficient of glucose in barrier layer b.

The transport through the cell layer (barrier layer c) is, according to Fig. 11.11C,

$$(D_B^g)_c c \frac{d^2 X_B^g}{dy^2} - \frac{d(J_B^g + J_B^e + J_A) X_B^g}{dy} + r_g = 0 \tag{11.54}$$

where $\left(D_B^g\right)_c$ is the diffusivity of glucose (m^2/s) in barrier layer c. Again by using

$J_A \gg J_B^g + J_B^e$, Eq. (11.54) becomes

$$\left(D_B^g\right)_c \frac{d^2 c_B^g}{dy^2} - \nu \frac{dc_B^g}{dy} + r_g = 0 \tag{11.55}$$

The solution of the differential Eq. (11.55) is

$$c_B^g = C_1^g exp\left(\frac{\nu}{\left(D_B^g\right)_c} y\right) + \left(\frac{r_g}{\nu} y\right) + C_2^g \tag{11.56}$$

where C_1^g and C_2^g are integral constants, which can be obtained from the boundary conditions;

Boundary condition (1):

$$c_B^g = c_{B3}^g \quad \text{at } y = 0 \tag{11.57}$$

Boundary condition (2): At the interface 3 ($y = 0$) glucose flux is also given by

$$J_B^g = -\left(D_B^g\right)_c \frac{dc_B^g}{dy} + \nu c_B^g \tag{11.58}$$

The molar flux at interface 4 is $J_B^g + (r^g V)/S$, as mentioned earlier, and it should be equal to the glucose flux through RO membrane, therefore,

$$J_B^g + \frac{r^g V}{S} = -\left(D_B^g\right)_c \frac{dc_B^g}{dy} + \nu c_B^g + \frac{r^g V}{S} = B_d^g\left(c_{B4}^g - c_{B5}^g\right) \tag{11.59}$$

where B_d^g is the glucose permeation constant of the RO membrane (barrier layer d).

From Eqs. (11.56) and (11.59)

$$C_2^g = \frac{B_d^g}{\nu}\left(c_{B4}^g - c_{B5}^g\right) - \frac{r^g V}{\nu S} + \frac{\left(D_B^g\right)_c r^g}{\nu^2} \tag{11.60}$$

(From Eq. (11.56) $c_B^g = C_1^g + C_2^g$ at $y = 0$. As well, $\frac{dc_B^g}{dy} = C_1^g \frac{\nu}{(D_B^g)_c} \exp(\frac{\nu}{(D_B^g)_c} y) + \frac{r_g}{\nu}$ and $\frac{dc_B^g}{dy} = \frac{C_1^g \nu}{(D_B^g)_c} + \frac{r_g}{\nu}$ at $y = 0$. Then, from Eqs. (11.59), (11.60) can be derived.)

Furthermore,

$$\frac{V}{S} = \delta_c \tag{11.61}$$

where δ_c is the thickness of the barrier layer c. Defining the mass transfer coefficient of glucose in barrier layer c as

$$\frac{\left(D_B^g\right)_c}{\delta_c} = k_c^g \tag{11.62}$$

$$C_2^g = \frac{B_d^g}{\nu}\left(c_{B4}^g - c_{B5}^g\right) - \frac{r^g V}{\nu S} + \frac{k_c^g r^g V}{\nu^2 S} \tag{11.63}$$

From Eqs. (11.56) and (11.57)

$$C_1^g + C_2^g = c_{B3}^g \tag{11.64}$$

Eqs. (11.63) and (11.64) allow calculation of the integration constants C_1^g and C_2^g.

Furthermore,

$$c_B^g = c_{B4}^g \text{ at } y = \delta_c \tag{11.65}$$

Therefore

$$c_{B4}^g = C_1^g \exp\left(\frac{\nu}{\left(D_B^g\right)_c}\delta_c\right) + \frac{r^g}{\nu}\delta_c + C_2^g \tag{11.66}$$

By using Eqs. (11.61) and (11.62)

$$c_{B4}^g = C_1^g \exp\left(\frac{\nu}{k_c^g}\right) + \frac{r^g V}{\nu S} + C_2^g \tag{11.67}$$

As for ethanol, it is assumed that ethanol permeates the membrane freely. Then, all the ethanol produced by the yeast cell goes into the permeate, therefore

$$c_{B5}^e = \frac{r^e V}{\nu S} \tag{11.68}$$

The rates of the biocatalytic reactions are:

Assuming 1 g of glucose is consumed for producing 20×10^{10} yeast cells, the glucose consumption is given by,

$$r^g = \frac{-\frac{1}{Y_{\frac{X}{S}}}\frac{dX}{dt}\left(1 - \frac{c_{B5}^e}{c_{Binhi}^e}\right) - mX}{20 \times 10^{10}} \tag{11.69}$$

where $\left(1 - c_{B5}^e / c_{Binhi}^e\right)$ shows the inhibition of yeast growth by the product ethanol.

For ethanol production,

$$r^e = -\frac{1}{Y_{\frac{P}{S}}} r^g \tag{11.70}$$

where $X, Y_{X/S}$, and $Y_{P/S}$ are, respectively, total cell number, cell mass yield coefficient, and product yield coefficient. The simulation algorithm is given in Scheme 11.6.

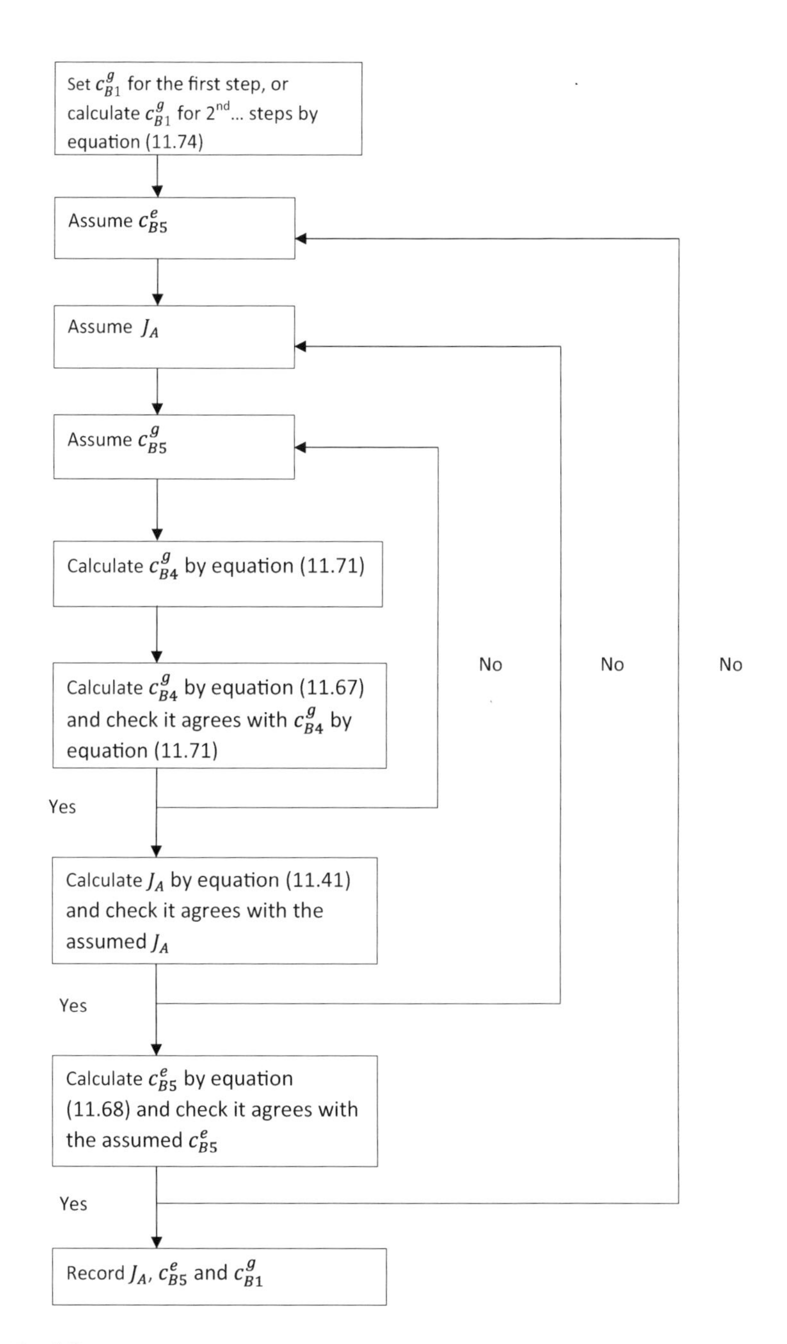

Scheme 11.6 Membrane reactor.

Problem 11.7:

Solve J_A, c_{B5}^{e}, and c_{B1}^{g} with the design parameters given in Table 11.1. The number of yeast cells increased from 1×10^9 to 14×10^9 in 210 h.

Table 11.1 Design parameters for Problem 11.7.

Parameters	Values	Units
A_b	3.239×10^{-6}	mol/m^2 s Pa
A_c	2.770×10^{-8}	mol/m^2 s Pa
A_d	1.750×10^{-7}	mol/m^2 s Pa
p_2	2859.3	kPa
p_5	101.352	kPa
B^g	1.429×10^5	kPa
B_d^g	1.280×10^{-8}	m/s
k_a^g	1.33×10^{-5}	m/s
k_b^g	1.16×10^{-6}	m/s
k_c^g	3.04×10^{-4}	m/s
S	6×10^{-4}	m^2
V	2×10^{-6}	m^3
$V_{\text{feed tank}}$	3×10^{-3}	m^3
$Y_{P/S}$	0.375	—
$Y_{X/S}$	0.06	—
m	2.778×10^{-4}	1 s^{-1}
$c^e_{B,\text{inhi}}$	1960	mol/m^3

Answer:

First assume $J_A = 1.63 \times 10^{-2}$ mol/m^2 s
Then, using Eq. (11.43)

$$\nu = \frac{1.63 \times 10^{-2}}{5.53 \times 10^4} = 2.948 \times 10^{-7} \text{ m/s}$$

For $t = 0$

1. Since the number of cells increased from 1×10^9 to 14×10^9 from $t = 0$ to $t = 210$ h

$$X(t = 0) = 10^9$$

$$\frac{dX}{dt} = \frac{(14 - 1) \times 10^9}{210 \times 3600} = 1.72 \times 10^4 1/\text{s}$$

First guess $c^e_{B5} = 776$ mol/m^3 is made.

Then, according to Eq. (11.69)

$$r^g = \frac{-\frac{1.72\times 10^4}{0.06}\left(1-\frac{776}{1960}\right) - 2.778\times 10^{-4}\times 10^9}{20\times 10^{10}\times 2\times 10^{-6}} = -1.127 \text{ g/m}^3 \text{ s}$$

$$\text{or} = -6.263\times 10^{-3} \text{ mol/m}^3 \text{ s}$$

Further, from Eq. (11.70)

$$r^e = -\frac{1}{0.375}\times(-1.127) = 3.01 \text{ g/m}^3 \text{ s}$$

$$\text{or} = 6.543\times 10^{-2} \text{ mol/m}^3 \text{ s}$$

From Eq. (11.42)

$$X_{B5}^g = \frac{J_B^g + \frac{r^g V}{S}}{J_A}$$

Hence,

$$J_B^g = J_A X_{B5}^g - \frac{r^g V}{S}$$

Using Eq. (11.43)

$$J_A = c\nu$$

Hence,

$$J_B^g = c\nu X_{B5}^g - \frac{r^g V}{S} = \nu c_{B5}^g - \frac{r^g V}{S}$$

Assume $c_{B5}^g = 3.5\times 10 \text{ mol/m}^3$, then

$$J_B^g = 2.947\times 10^{-7}\times 3.5\times 10 - \frac{-6.263\times 10^{-3}\times 2\times 10^{-6}}{6\times 10^{-4}} = 3.119\times 10^{-5} \text{ mol/m}^2 \text{ s}$$

At the RO membrane (barrier layer *d*)
$B_d^g\left(c_{B4}^g - c_{B5}^g\right) = \nu c_{B5}^g$, hence,

$$c_{B4}^g = \left(\nu + B_d^g\right)\frac{c_{B5}^g}{B_d^g} \tag{11.71}$$

Inserting numerical values,

$$c_{B4}^g = \left(2.948\times 10^{-7} + 1.280\times 10^{-8}\right)\times 3.5\times 10/1.280\times 10^{-8} = 841.1 \text{ mol/m}^3 \tag{11.72}$$

From Eq. (11.51)

$$c_{B2}^{g} = \left(c_{B1}^{g} - c_{B5}^{g} + \left[\frac{r^{g}V/S}{\nu}\right]\right)e^{\frac{\nu}{k_{a}^{g}}} + c_{B5}^{g} - \left[\frac{r^{g}V/S}{\nu}\right]$$

Inserting numerical values,

$$c_{B2}^{g} = \left(6.67\times10^{2} - 3.5\times10 + \frac{-6.263\times10^{-3}\times2\times10^{-6}}{2.948\times10^{-7}\times6\times10^{-4}}\right)$$
$$\times\exp\left(\frac{2.948\times10^{-7}}{1.33\times10^{-5}}\right) + 3.5\times10 - \frac{-6.263\times10^{-3}\times2\times10^{-6}}{2.948\times10^{-7}\times6\times10^{-4}}$$
$$= 679.3\ \text{mol/m}^{3}$$

From Eq. (11.53)

$$c_{B3}^{g} = \left(6.67\times10^{2} - 3.5\times10 + \frac{-6.263\times10^{-3}\times2\times10^{-6}}{2.948\times10^{-7}\times6\times10^{-4}}\right)$$
$$\times\exp\left(\frac{2.948\times10^{-7}}{1.16\times10^{-6}}\right) + 3.5\times10 - \frac{-6.263\times10^{-3}\times2\times10^{-6}}{2.948\times10^{-7}\times6\times10^{-4}}$$
$$= 829.2\ \text{mol/m}^{3}$$

From Eq. (11.63)

$$C_{2}^{g} = \frac{1.280\times10^{-8}}{2.948\times10^{-7}}(841.1 - 3.5\times10) - \frac{-6.263\times10^{-3}\times2\times10^{-6}}{2.948\times10^{-7}\times6\times10^{-4}}$$
$$+ \frac{-6.263\times10^{-3}\times2\times10^{-6}}{2.948\times10^{-7}\times6\times10^{-4}}\times\frac{3.04\times10^{-4}}{2.948\times10^{-7}} = -72924$$

From Eq. (11.64)

$$C_{1}^{g} = c_{B3}^{g} - C_{2}^{g} = 829.2 - (-72924) = 73753$$

From Eq. (11.67)

$$c_{B4}^{g} = 73753\times\exp\left(\frac{2.948\times10^{-7}}{3.04\times10^{-4}}\right) + \frac{-6.263\times10^{-3}\times2\times10^{-6}}{2.948\times10^{-7}\times6\times10^{-4}} + 72924 = 830.28 \quad (11.73)$$

c_{B4}^{g} obtained from Eqs. (11.72) and (11.73) are different. This is because the assumption of $c_{B5}^{g} = 3.5\times10\ \text{mol/m}^{3}$ was wrong.
When $c_{B5}^{g} = 3.455\times10\ \text{mol/m}$ c_{B4}^{g} is831 from both Eqs. (11.72) and (11.73).
From Eq. (11.39)

$$p_{3} = 2,859,300 - \frac{0.0163}{3.239\times10^{-6}} = 2,854,268\ \text{Pa}$$

From Eq. (11.40)

$$p_4 = 2,854,268 - \frac{0.0163}{2.770 \times 10^{-8}} = 2,265,820$$

From Eq. (11.41)

$$J_A = 1.750 \times 10^{-7}\left(2,265,820 - 101,325 - 1.429 \times 10^8 \times \frac{830.28}{5.53 \times 10^4} + 1.429 \times 10^8 \times \frac{34.55}{5.53 \times 10^4}\right) = 0.01895 \text{ mol/m}^2\text{ s}$$

which does not agree with the first guess of $J_A = 1.63 \times 10^{-2}$ mol/m^2 s.

$J_A = 1.65 \times 10^{-2}$ mol/m^2 s will satisfy both the first guess and the J_A by Eq. (11.41).

Corresponding to this J_A,

c_{B5}^g is amended to 34.40 to make c_{B5}^g values from Eqs. (11.72) and (11.73) agree.

According to Eq. (11.68)

$$c_{B5}^e = \frac{6.543 \times 10^{-2} \times 2 \times 10^{-6}}{2.948 \times 10^{-7} \times 6 \times 10^{-4}} = 739.8 \text{ mol/m}^3$$

which does not agree with the guessed 776 mol/m^3.

The initial guess of 740 mol/m^3 makes c_{B5}^e obtained from Eq. (11.68) agree with the guessed value. Under this c_{B5}^e, there is no change in J_A and c_{B5}^g.

2. For the second 20 h

The solution volume in the feed tank decreased in the first 20 h to

$$V_{feedtank} = 3 \times 10^{-3} - \frac{\left(18.02 \times 1.65 \times 10^{-2} \times 3600 \times 20 \times 6 \times 10^{-4}\right)}{10^6} = 0.002987$$

The feed glucose concentration will be

$$c_{B1}^g = \frac{\left(3 \times 10^{-3} \times 6.67 \times 10^2 - 3.14 \times 10^{-5} \times 3600 \times 20 \times 6 \times 10^{-4}\right)}{0.002987} = 669.4 \tag{11.74}$$

Using the above c_{B1}^g, the computation is repeated for the second 20 h.

The calculation is repeated up to 140 h and the results, permeation rate (SJ_A), glucose concentration in feed (c_{B1}^g), and ethanol concentration in permeate (c_{B5}^e) are reported in Fig. 11.12.

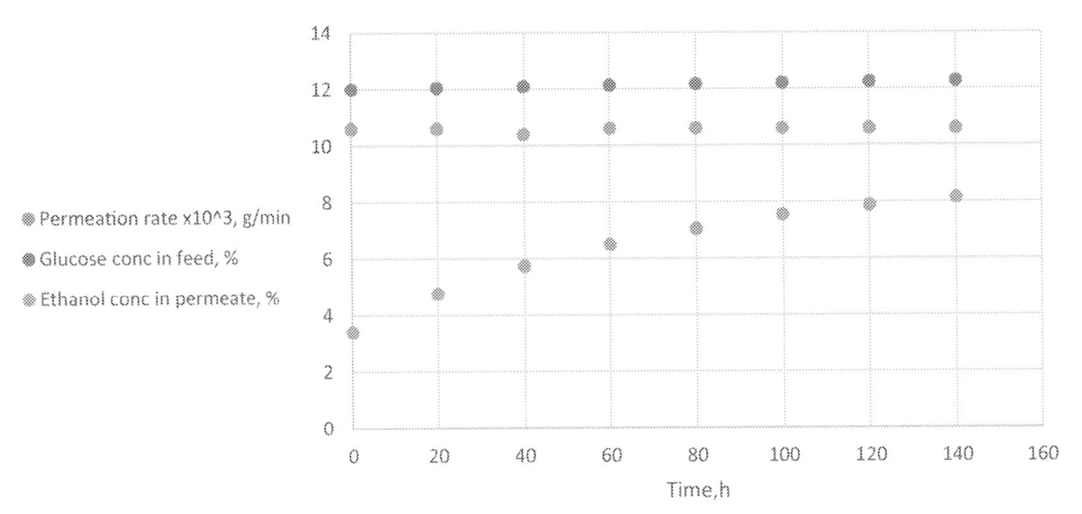

Figure 11.12 Results of the bioreactor calculation, SJ_A, c_{B1}^{g} and c_{B5}^{e} versus t.

Nomenclature

Symbol	Definition [dimension (SI unit)]
A	Pure water permeation coefficient (m^3/m^2 s Pa)
b	Osmotic pressure coefficient (Pa m^3/mol)
B	Solute permeation constant (m/s)
c_i	Inlet solute concentration (mol/m^3)
c_f	Feed solute concentration (mol/m^3)
c_p	Permeate solute concentration (mol/m^3)
c_{po}	Solute concentration of the permeate at the system outlet (mol/m^3)
c_R	Retentate solute concentration (mol/m^3)
c_{RO}	Solute concentration of the retentate at the system outlet (mol/m^3)
J_A	Flux of water (m^3/m^2 s)
J_B	Flux of solute (mol/m^2 s)
S	Membrane area (m^2)
u_i	Inlet flow rate (m^3/s)
u_f	Feed flow rate (m^3/s)
u_p	Permeate flow rate (m^3/s)
u_{po}	Permeate flow rate at the system outlet (m^3/s)
u_R	Retentate flow rate (m^3/s)
u_{RO}	Retentate flow rate at the system outlet (m^3/s)
x	Recycle ratio
Greek letter	
Δp	Cross-membrane pressure difference (Pa)
Superscripts	
1	First stage
2	Second stage

d_e	Effective diameter of hollow fiber (m)
d_i	Inner diameter of hollow fiber (m)
J_A	Flux of component A (mol/m^2 s)
J_B	Flux of component B (mol/m^2 s)
$\boldsymbol{L}$	Length of hollow fiber (m)
p_1	Pressure on the feed side (Pa)
p_3	Pressure on the permeate side (Pa)
$\frac{P_A}{\delta}$	Permeance of component A (mol/m^2 s Pa)
$\frac{P_B}{\delta}$	Permeance of component B (mol/m^2 s Pa)
S_1	Total cross-sectional area of the space between hollow fibers (m^2)
S_3	Total lumen cross-sectional area of follow fibers (m^2)
$\boldsymbol{u}_1$ and $\boldsymbol{u}_f$	Molar flow rate of feed gas mixture per unit cross-sectional area of the space between hollow fibers (mol/m^2 s)
$\boldsymbol{u}_3$ and $\boldsymbol{u}_p$	Molar flow rate of permeate gas mixture per unit area of hollow fiber cross-section (mol/m^2 s)
$\boldsymbol{u}_i$	Molar flow rate of feed gas mixture per unit cross-sectional area of the space between hollow fibers at the module inlet (mol/m^2 s)
$\boldsymbol{u}_{po}$	Molar flow rate of permeate gas mixture per unit area of hollow fiber cross-section at the system outlet (mol/m^2 s)
$\boldsymbol{u}_R$	Molar flow rate of retentate gas mixture per unit cross-sectional area of the space between hollow fibers (mol/m^2 s)
$\boldsymbol{u}_{RO}$	Molar flow rate of retentate gas mixture per unit cross-sectional area of the space between hollow fibers at the system outlet (mol/m^2 s)
$\boldsymbol{x}$	Recycle ratio
X_{A1} and X_{Af}	Mole fraction of component A in the feed gas mixture
X_{A3} and X_{Ap}	Mole fraction of component A in the gas mixture permeating through the membrane
$\overline{X}_{A3}$ and $\overline{X}_{Ap}$	Mole fraction of component A flowing on the lumen side of membrane
X_{Ai}	Mole fraction of component A at the module inlet
X_{Apo}	Mole fraction of component A in the gas mixture at the system outlet
X_{AR}	Mole fraction of component A in the retentate gas mixture
X_{ARO}	Mole fraction of component A in the retentate gas mixture at the system outlet
Greek letter	
μ	Viscosity of gas mixture (Pa s)
A_{FO}	Pure water permeation coefficient of FO membrane (m^3/m^2 s Pa)
A_{RO}	Pure water permeation coefficient of RO membrane (m^3/m^2 s Pa)
b	Osmotic pressure coefficient (Pa m^3/mol)
B_{FO}	Solute permeation constant of FO membrane (m/s)
B_{RO}	Solute permeation constant of RO membrane (m/s)
$c_{AL/SL}$	Solute concentration at the active layer/support layer interface (mol/m^3)
c_d	Solute (sodium chloride) concentration of the DS in FO (mol/m^3)
c_{DS}	Solute (sodium chloride) concentration of the DS in FO in equation (3.55) (mol/m^3)
c_f	Solute (sodium chloride) concentration of the feed solution in FO (mol/m^3)

$c_{O,FO}$	Solute concentration of FO outlet (mol/m^3)
$c_{p,FO}$	Solute concentration of FO permeate (mol/m^3)
$c_{p,RO}$	Solute concentration of RO permeate (mol/m^3)
$c_{R,RO}$	Solute concentration of RO retentate (mol/m^3)
$J_{A,FO}$	Volumetric flux of water in FO (m^3/m^2 s)
$J_{A,RO}$	Volumetric flux of water in RO (m^3/m^2 s)
$J_{B,FO}$	Solute flux in FO (mol/m^2 s)
k_m	Mass transfer coefficient (m/s)
$\boldsymbol{S}$	Membrane area (m^2)
u_d	Flow rate of the DS to FO (m^3/s)
u_f	Flow rate of the feed solution to FO (m^3/s)
$u_{O,FO}$	Flow rate o FO outlet (m^3/s)
$u_{p,FO}$	Flow rate of FO permeate (m^3/s)
$u_{p,RO}$	Flow rate of RO permeate (m^3/s)
$u_{R,RO}$	Flow rate of RO retentate (m^3/s)
Greek letters	
Δp	Pressure applied at RO (Pa gauge)
$\pi_{AL/SL}$	Osmotic pressure at the active layer/support layer interface (Pa)
π_{DS}	Osmotic pressure of DS (Pa)
π_{FS}	Osmotic pressure of feed solution (Pa)

$\boldsymbol{A}$	Pure water permeation coefficient (mol/m^2 s Pa)
c	Total molar concentration (mol/m^3)
c_B	Concentration of ethanol or glucose (mol/m^3)
c^e_{Binhi}	Inhibition concentration of ethanol (mol/m^3)
C^g_1	First integral constant (mol/m^3)
C^g_2	Second integral constant (mol/m^3)
D^g_B	Diffusivity of glucose (m^2/s)
J_A	Molar flux of water (mol/m^2 s)
J_B	Molar flux of glucose or ethanol (mol/m^2 s)
$\boldsymbol{k}$	Mass transfer coefficient (m/s)
m	Cell maintenance coefficient (1/s)
$\boldsymbol{p}$	Pressure (Pa)
r^e	Ethanol production rate (mol/m^3 s)
r^g	Glucose consumption rate (mol/m^3 s)
S	Effective area of cell layer (m)
$\boldsymbol{t}$	Time (s)
ν	Permeation velocity through barrier layers (m)
V	Effective volume of cell layer (m^3)
$\boldsymbol{V}_{\textbf{feed tank}}$	Volume of feed tank (m^3)
X	Total cell number
X^g_B	Mole fraction of glucose
$Y_{\frac{P}{S}}$	Product yield coefficient
$Y_{\frac{X}{S}}$	Cell mass yield coefficient
Greek letters	
δ	Barrier layer thickness (m)
π	Osmotic pressure (Pa)
Superscripts	
e	ethanol
g	glucose
Subscripts	
1	feed
2, 3, 4	Barrier boundaries

5	Permeate
a, b, c, d	Barrier layers
A	Water
B	Ethanol or glucose

References

Altaee, A., Hilal, N., 2015. High recovery rate NF–FO–RO hybrid system for inland brackish water treatment. Desalination 363, 19–25.

Cath, T.Y., Hancock, N.T., Lundin, C.D., Hoppe-Jones, C., Drewesetal, J.E., 2010. A multi-barrier osmotic dilution process for simultaneous desalination and purification of impaired water. J. Membr. Sci 362, 417–426.

Hancock, N.T., Black, N.D., Cath, T.Y., 2012. A comparative lifecycle assessment of hybrid osmotic dilution desalination and established seawater desalination and wastewater reclamation processes. Wat. Res 46, 1145–1154.

Jeong, Y.S., Vieth, W.R., Matsuura, T., 1989. Study of transport phenomena in an immobilized yeast membrane bioreactor. Ind. Eng. Chem. Res. 28, 231.

Matsuura, T., 1994. Synthetic Membranes and Membrane Separation Processes. CRC Press, Florida, Chapter 9.

Mulder, M.H.V., Smolders, C.A., 1966. Pervaporation in continuous alcohol fermentation. In: Bakish, R. (Ed.), Proceedings of the First International Conference on Pervaporation Processes in the Chemical Industry. Bakish Materials Corp, Englewood, NJ, p. 187.

Vasdevan, M., Matsuura, T., Chotani, G.K., Vieth, W.R., 1987. Simultaneous bioreaction and separation of an immobilized yeast membrane reactor. Sep. Sci. Technol 22, 1651.

Yaeli, J., 1992. Method and apparatus for processing liquid solutions of suspensions particularly useful in the desalination of saline water. Google patents.

12

Cost of water

12.1 Calculation of water production cost by Desalination Economic Evaluation Program 2000

The economic evaluation of membrane processes is difficult to deal with for many scientists and engineers as they have been educated within different disciplines. Nevertheless, we cannot avoid this topic in many cases. In this chapter, the cost of drinking water production is evaluated according to the Desalination Economic Evaluation Program (DEEP) of the International Atomic Energy Agency (IAEA). The software of DEEP is revised every few years, with the newest versions being DEEP-3 and -5, released in 2006 and 2013, respectively (Desalination Economic Evaluation Program DEEP User's Manual DEEP-3.0, 2006; DEEP 5 User Manual, 2013). This program was created to calculate the production cost of water when various desalination processes are combined with various power sources. It, however, allows the cost evaluation of water production for the stand-alone reverse osmosis (RO) process, when electricity is purchased.

Although DEEP has progressed with every new version, we visit the oldest version of DEEP that was released in 2000 (Desalination Economic Evaluation Program DEEP User's Manual, 2000), as the process of water cost calculation is described in most detail in this version, and an example is given to calculate the cost for a model case. In the following, the "row" number is adopted from DEEP, "parameter" explains what the item is, "unit" is the unit of the item, and the "variable name" is the code given for the item. The algorithm of the calculation is given in Scheme 12.1.

Membrane Separation Processes. DOI: https://doi.org/10.1016/B978-0-12-819626-7.00001-6

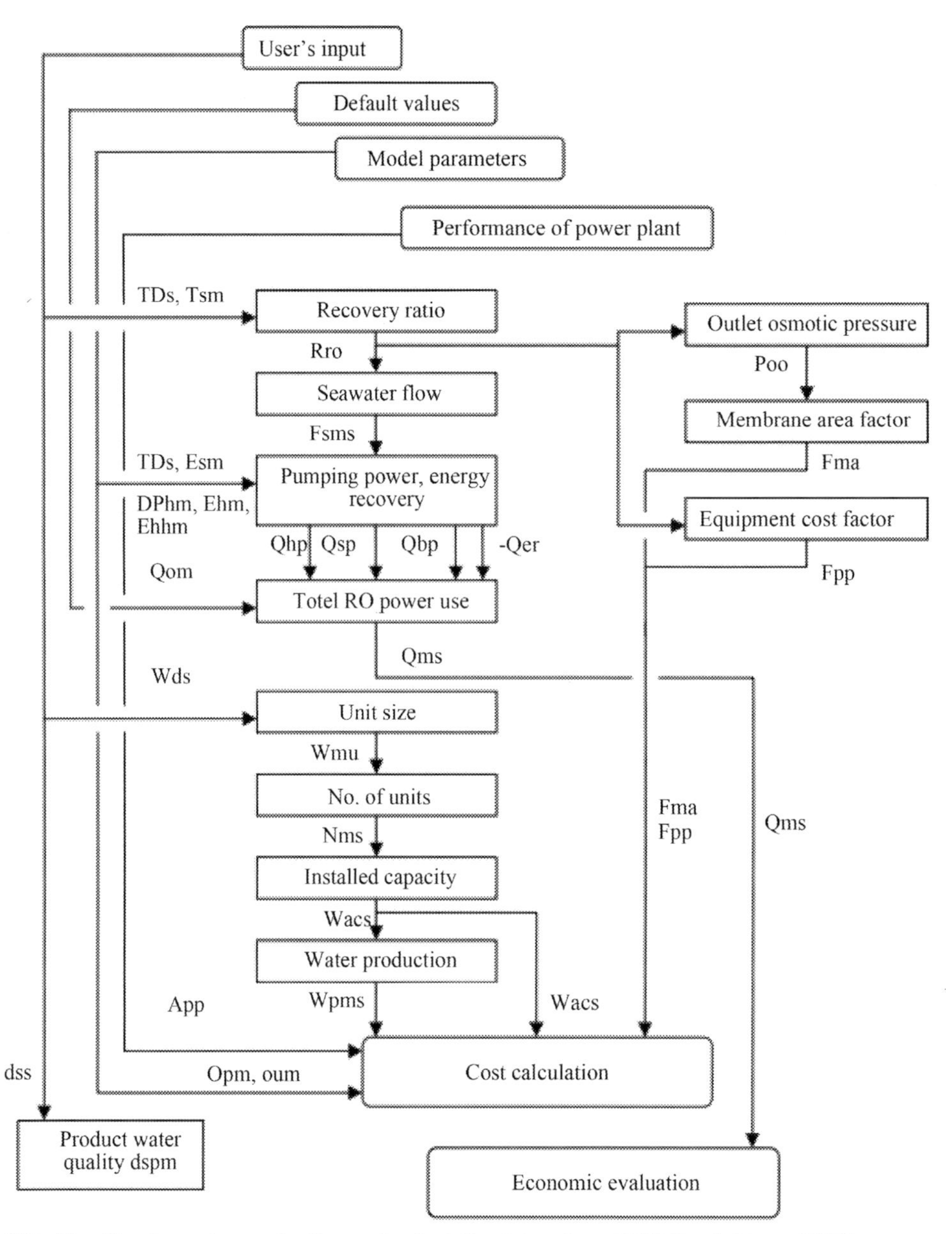

Scheme 12.1 Algorithm for water production cost calculation taken from DEEP User's Manual 2000.

12.1.1 Case identification and site characteristics box

1. Row (10), parameter (required water plant capacity at site), unit [$/(m^3/d)], variable name (Wds)
   100,000
2. Row (17), parameter (membrane type), unit (text), variable name (MemType)
   Spiral wound (SW)

12.1.2 Technical parameter input data

1. Row (25), parameter (average annual cooling water temperature), Unit (°C), variable name (Tsw)
   Annual average seawater temperature at site, 30
2. Row (26), parameter (seawater total dissolved solid), unit (ppm), variable name (TDS)
   35,000
3. Row (27), parameter (electric motor efficiency), unit (–), variable name (Eem)
   0.96
4. Row (51), parameter (planned outage time), unit (–), variable name (opp)
   The average planned downtime in days divided by 365, typically 0.1 for a power plant and 0.05 for a heat-only plant, this is not applicable in the case of no power generation in the plant
5. Row (52), parameter (unplanned outage time), unit (–), variable name (oup)

The average unplanned downtime in days divided by 365, typically 0.11 for a power plant and 0.15 for a heat-only plant, this is not applicable in the case of no power generation in the plant

12.1.3 Membrane water performance data

1. Row (84), parameter (seawater pump head), unit (bar), variable name (DPsm)
   Typically, 1.7
2. Row (85), parameter (seawater pump efficiency), unit (–), variable name (Esm)
   Usually, 0.70–0.85, average is 0.775
3. Row (88), parameter (booster pump head), unit (bar), variable name (Dpbm)
   Typically, 3.3
4. Row (89), parameter (booster pump efficiency), unit (–), variable name (Ebm)
   Usually 0.70–0.85, average is 0.775

Table 12.1 Recovery ratio of spiral wound module.

Temperature (°C)	Feed salt concentration (ppm)			
	35,000	38,000	43,000	45,000
15	62.7	65.3	67.7	68.4
20	64.1	66.4	68.3	68.8
25	65.9	68.0	68.8	69.2
30	62.2	65.3	66.4	68.5
35	58.2	62.4	67.8	68.8
40	55.0	59.1	69.0	69.2

5. Row (90), parameter (high head pump pressure rise), unit (bar), variable name (DPhm)

   From Table 12.1 for SW, 62.2
6. Row (91), parameter (high head pump efficiency), unit (–), variable name (Ehm)

   Usually, 0.7–0.85, average is 0.775
7. Row (92), parameter (hydraulic pump hydraulic coupling efficiency), unit (–), variable name (Ehhm)

   Typically, 0.97
8. Row (93), parameter (energy recovery efficiency), unit (–), variable name (Ehm)

   Usually, 0.7–0.9, average is 0.8
9. Row (94), parameter (other specific power use), unit [kW(e) h/m^3], variable name (Qsom)

   Remaining specific electric power use

   Typically, 0.0408 or 0.98, 0.98 is taken
10. Row (95), parameter (planned outage rate), unit (–), variable name (opm)

    Average planned downtime in days divided by 365, typically, 0.032

    Row (96), parameter (unplanned outage rate), unit (–), variable name (oum)
11. Average unplanned downtime in days divided by 365, typically, 0.06

12.1.4 Economic parameters input data

1. Row (103), parameter (discount rate), unit (%/a), variable name (i)

   Typically, 8% for developing countries

2. Row (104), parameter (interest rate), unit (%/a), variable name (ir)
   Typically, 8%
3. Row (107), parameter (plant economic life), unit (a), variable name (Lep)
   The power/water plant life time used in the cost calculations, 10
4. Row (108), parameter (purchased electricity cost), unit [$/kW(e)h], variable name (Cpe)
   0.06

12.1.5 Reverse osmosis plant cost input data

1. Row (142), parameter (base unit cost), unit [$/(m^3/d)], variable name (Cmu)
   The specific overnight costs of the membrane plant are typically 800 $/(m^3/d) for 24,000 m^3/d unit size
2. Row (143), parameter (optional in/outfall specific base cost), unit [$/(m^3/d)], variable name (Csmo)
   If not specified, the value has to be set to 0
3. Row (144), parameter (ratio membrane eq. cost to total cost), unit (–), variable name (kme)
   Typically 0.20
4. Row (145), parameter (water plant cost contingency factor), unit (–), variable name (kmc)
   Typically 0.1–0.20, average is 0.15
5. Row (146), parameter (water plant owner's cost factor), unit (–), variable name (kmo)
   Typically 0.05
6. Row (147), parameter (water plant lead time), unit (m), variable name (Lm)

$$\text{Formula} = \left(IF\left(\left(\text{INT}\left(\frac{\text{Wds}}{8\times 24{,}000}\right)+1\right)\times 12\right)\right.$$

$$\left.<36;\left(\text{INT}\left(\frac{\text{Wds}}{8\times 24{,}000}\right)+1\right)\times 12;36\right)$$

Since

$$\text{Wds} = 100{,}000,\ \left(\text{INT}\left(\frac{100{,}000}{8\times 24{,}000}\right)+1\right)\times 12 = 24,$$

7. Row (148), parameter (average management salary), unit ($/a), variable name (Smm)
   200,000

8. Row (149), parameter (average labor salary), unit ($/a), variable name (Sml)
   100,000
9. Row (150), parameter (O&M membrane replacement cost), unit ($/m^3), variable name (cmm)
   Typically 0.06
10. Row (151), parameter (O&M spare parts cost), unit ($/m^3), variable name (cmsp)
    Typically 0.04
11. Row (152), parameter (specific chemicals cost for pretreatment), unit ($/m^3), variable name (cmcpr)
    Typically 0.03
12. Row (153), parameter (specific chemicals cost for post-treatment), unit ($/m^3), variable name (cmcpo)
    Typically 0.03
13. Row (154), parameter (water plant A&M insurance cost), unit (%), variable name (kmi)
    Typically, 0.5% of the base capital costs
14. Row (163), parameter (total site-specific base power plant net output), unit [MW(e)], variable name (Pen)

Not applicable in the case of no power generation in the plant

12.1.6 Single-purpose plant performance

1. Row (175), parameter (operating availability), unit (–), variable name (App)

$$\text{Formula} = (1 - \text{opp}) \times (1 - \text{oup})$$

This is not applicable in the case of no power generation in the plant

12.1.7 Stand-alone reverse osmosis water plant performance

1. Row (219), parameter (required water plant production capacity), unit (m^3/d), variable name (Wcst)

$$\text{Formula} = \text{Wds} = 100{,}000$$

2. Row (220), parameter (RO feed inlet temperature), unit (°C), variable name (Tim)
   seawater temperature, Tws is used, 30

Table 12.2 Recovery ratio of spiral wound membranes.

Temperature (°C)	Feed salt concentration (ppm)			
	35,000	38,000	43,000	45,000
15	0.411	0.401	0.383	0.374
20	0.411	0.401	0.383	0.374
25	0.411	0.401	0.383	0.374
30	0.401	0.391	0.383	0.374
35	0.390	0.382	0.372	0.363
40	0.378	0.372	0.361	0.353

3. Row (221), parameter (approximate inlet osmotic pressure), unit (bar), variable name (Pio)

$$\text{Formula} = 0.0000348 \times (\text{Tim} + 273) \times \frac{\text{TDS}}{14.7}$$

$$= 0.0000348 \times (30 + 273) \times \frac{35{,}000}{14.7} = 25.11$$

4. Row (222), parameter (recovery ratio), unit (–), variable name (Rro)

$$\text{Formula} = \text{memrec}(\text{MemType}; \text{TDS}; \text{Tim}) = \text{memrec}(\text{SW}; 35,000; 30),$$

see Table 12.2, 0.401.

5. Row (223), parameter (outlet dissolved solids concentration), unit (ppm), variable name (dso)

$$\text{Formula} = \frac{\text{TDS}}{(1 - Rro)} = \frac{35{,}000}{(1 - 0.401)} = 58,431$$

6. Row (224), parameter (approximate outlet osmotic pressure), unit (bar), variable name (Poo)

$$\text{Formula} = 0.0000348 \times (\text{Tim} + 273) \times \frac{\text{dso}}{14.7}$$

$$= 0.0000348 \times (30 + 273) \times \frac{58,431}{14.7} = 41.91$$

7. Row (225), parameter (membrane area factor), unit (–), variable name (Fma)

$$\text{memarea}(\text{MemType}; \text{TDS}; \text{Tim}) = \text{memarea}(\text{SW}; 35{,}000; 30) = 230,$$

Table 12.3 Number of spiral wound membrane elements including six vessels.

Temperature (°C)	Feed salt concentration (ppm)			
	35,000	38,000	43,000	45,000
15	340	360	420	430
20	280	300	360	385
25	228	240	310	340
30	230	240	290	315
35	240	245	258	282
40	250	255	227	255

see Number of Vessel in Table 12.3.

$$\text{Formula} = \frac{\text{memarea(MemType; TDS; Tim)}}{228} = \frac{230}{228} = 1.01$$

8. Row (226), parameter (pretreatment, pump and piping size increase factor), unit (–), variable name (Fpp)

$$\text{Formula} = IF\left(\text{Mem Type} = \text{HF}; \left(\frac{0.5}{\text{Pro}}\right)^{0.5}; \left(\frac{0.411}{\text{Pro}}\right)^{0.5}\right) = \left(\frac{0.411}{0.401}\right)^{0.5} = 1.012$$

9. Row (227), parameter (product water quality before pretreatment), unit (ppm), variable name (dspms)

$$\text{Formula} = \text{TDS} \times 0.0078 = 35{,}000 \times 0.0078 = 273$$

10. Row (228), parameter (default unit size), unit (m^3/d), variable name (Wmud)

$$\text{Formula} = 24{,}000 \text{ for Wcst} \leq 100{,}000$$

11. Row (229), parameter (selected unit size), unit (m^3/d), variable name (Wmu)

$$\text{Formula} = \text{Wmud} = 24{,}000$$

12. Row (230), parameter (number of RO units), unit (–), variable name (Nms)

    Number of RO units with a capacity of Wmu to produce West

$$\text{Formula} = \text{INT}\left(\frac{\text{West}}{\text{Wmu}}\right) + 1 = \text{INT}\left(\frac{100{,}000}{24{,}000}\right) + 1 = 5$$

13. Row (231), parameter (installed stand-alone RO capacity), unit (m^3/d), variable name (Wacs)

The real installed RO capacity to enable the targeted water production of West

$$\text{Formula} = \text{Wmu} \times \text{Nms} = 24{,}000 \times 5 = 120{,}000$$

14. Row (232), parameter (seawater flow), unit (kg/s), variable name (Fsms)

Flow rate of seawater that enters the RO unit

$$\text{Formula} = \frac{\text{Wacs} \times 1000}{\text{Pro} \times 24 \times 3600} = \frac{120{,}000 \times 1000}{0.401 \times 24 \times 3600} = 3464$$

15. Row (233), parameter (stand-alone seawater pumping power), unit [MW(e)], variable name (Qsp)

Power required for seawater intake

$$\text{Formula} = \frac{\text{DPsm} \times \text{Fsms}}{10{,}000 \times \text{Esm} \times \text{Eem}} = \frac{1.7 \times 3464}{1000 \times 0.775 \times 0.96} = 0.792$$

16. Row (234), parameter (booster pump power), unit [MW(e)], variable name (Qbp)

Power required to run booster pump

$$\text{Formula} = \frac{\text{DPbm} \times \text{Fsms}}{10{,}000 \times \text{Ebm} \times \text{Eem}} = \frac{3.3 \times 3464}{10{,}000 \times 0.775 \times 0.96} = 1.536$$

17. Row (235), parameter (high head pump power), unit [MW(e)], variable name (Qhp)

Power required to run high-pressure pump

$$\text{Formula} = \frac{\text{Dphm} \times \text{Fsms}}{10{,}000 \times \text{Ehm} \times \text{Eem} \times \text{Ehhm}}$$
$$= \frac{62.2 \times 3464}{10{,}000 \times 0.775 \times 0.96 \times 0.97} = 29.86$$

18. Row (236), parameter (energy recovery), unit [MW(e)], variable name (Qer)

Energy recovery is modeled as a reverse running pump, mechanically coupled to the shaft of the high head pump

$$\text{Formula} = -\frac{\text{DPhm} \times \text{Fsms} \times (1 - \text{Rro}) \times \text{Eer}}{10{,}000}$$
$$= -\frac{62.2 \times 3464 \times (1 - 0.401) \times 0.8}{10{,}000} = -10.32$$

19. Row (237), parameter (other power), unit [MW(e)], variable name (Qom)

$$\text{Formula} = \frac{\text{Wacs} \times \text{Qsom}}{24 \times 1000} = \frac{120{,}000 \times 0.98}{24 \times 1000} = 4.90$$

20. Row (238), parameter (total stand-alone RO power use), unit [MW(e)], variable name (Qms)

$$\text{Formula} = \text{Qsp} + \text{Qbp} + \text{Qhp} + \text{Qer} + \text{Qom}$$
$$= 0.792 + 1.536 + 29.86 - 10.32 + 4.90 = 26.77$$

21. Row (239), parameter (membrane water plant availability), unit (–), variable name (Apm)

$$\text{Formula} = (1 - \text{opm}) \times (1 - \text{oum}) = (1 - 0.032) \times (1 - 0.06) = 0.910$$

22. Row (240), parameter (combined power/water plant load factor), unit (–), variable name (Acpm)

$$\text{Formula} = \text{App} \times (1 - \text{oum})$$

This is not applicable in the case of no power generation in the plant

23. Row (241), parameter (annual water production), unit (m^3/a), variable name (Wpms)

$$\text{Formula} = \text{Wacs} \times \text{Apm} \times 365 = 120{,}000 \times 0.910 \times 365 = 39{,}858{,}000$$

24. Row (242), parameter (average daily water production), unit (m^3/d), variable name (Wpmsad)

$$\text{Formula} = \text{Wacs} \times \text{Apm} = 120{,}000 \times 0.910 = 109{,}200$$

25. Row (243), parameter (specific stand-alone power consumption), unit [kW(e)h/m^3], variable name (Qcms)

$$\text{Formula} = \frac{24 \times 1000 \times \text{Qms}}{\text{Wacs}} = \frac{24 \times 1000 \times 26.77}{120{,}000} = 5.354$$

26. Row (244), parameter (net stand-alone saleable), unit [MW(e)], variable name (Qssp)
    Formula = Pen – Qms
    Not applicable in the case of no power generation in the plant

12.1.8 Stand-alone reverse osmosis water plant costs

1. Row (329), parameter (number of units), unit (–), variable name (Nms1)

$$\text{Formula} = \text{Nms} = 5$$

2. Row (330), parameter (correction factor for unit size), unit (–), variable name (kmsus)

$$\text{Formula} = \frac{1}{\left(\frac{\text{Wmu}}{24000}\right)^{0.15}} = \frac{1}{\left(\frac{24000}{24000}\right)^{0.15}} = 1$$

3. Row (331), parameter (correction factor for number of units), unit (–), variable name (kmsnu)

$$\text{Formula} = \frac{1}{\text{Nms1}^{0.1}} = \frac{1}{5^{0.1}} = 0.851$$

4. Row (332), parameter (correction factor for TDS and temperature), unit (–), variable name (kmstt)

$$\text{Formula} = \text{Fma} \times \text{kem} + \text{Fpp} \times (1 - \text{kem})$$
$$= 1.01 \times 0.2 + 1.012 \times (1 - 0.8) = 1.012$$

5. Row (333), parameter (adjusted water plant-specific cost), unit [$/(m^3/d)], variable name (Cmsa)

$$\text{Formula} = \text{Cmu} \times \text{kmsus} \times \text{kmsnu} \times \text{kmstt}$$
$$= 800 \times 1 \times 0.851 \times 1.012 = 689$$

6. Row (334), parameter (stand-alone in/outfall specific cost), unit [$/(m^3/d)], variable name (Cmio)

$$\text{Formula} = IF\left(\text{Csmo} > 0; \text{Csmo}; \frac{7{,}400{,}200 \times \left(\frac{\text{Fsms}}{486}\right)^{0.45}}{\text{Wacs}}\right)$$
$$= \frac{7{,}400{,}200 \times \left(\frac{3464}{486}\right)^{0.45}}{120{,}000} = 149.2$$

7. Row (335), parameter (stand-alone water plant-specific cost), unit [$/(m^3/d)], variable name (Cms)

$$\text{Formula} = \text{Cmsa} + \text{Cmio} = 689 + 149.2 = 838.2$$

8. Row (336), parameter (stand-alone water plant adjusted base cost), unit (M$), variable name (Cmsab)

$$\text{Formula} = \frac{\text{Wacs} \times \text{Cms}}{1{,}000{,}000} = \frac{120{,}000 \times 838.2}{1{,}000{,}000} = 100.58$$

9. Row (337), parameter (water plant owner's cost), unit (M$), variable name (DCmso)

$$\text{Formula} = \text{Cmsab} \times \text{kmo} = 100.58 \times 0.05 = 5.03$$

10. Row (338), parameter (water plant contingency cost), unit (M$), variable name (DCmsc)

$$\text{Formula} = (\text{Cmsab} + \text{DCmso}) \times \text{kmc} = (100.58 + 5.03) \times 0.15 = 15.84$$

11. Row (339), parameter (stand-alone water plant total construction cost), unit (M$), variable name (Cmscon)

$$\text{Formula} = \text{Cmsab} + \text{DCmso} + \text{DCmsc} = 100.58 + 5.03 + 15.84 = 121.45$$

12. Row (340), parameter (number of management personnel), unit (–), variable name (Nmsm)

$$\text{Formula} = \text{INT}\left(\frac{5 + \frac{\text{Wacs}}{55000}}{2}\right) = \text{INT}\left(\frac{5 + \frac{120000}{55000}}{2}\right) = 3$$

13. Row (341), parameter (O&M management cost), unit (M$/a), variable name (Cmsm)

$$\text{Formula} = \frac{\text{Nmsm} \times \text{Smm}}{1{,}000{,}000} = \frac{3 \times 200{,}000}{1{,}000{,}000} = 0.6$$

14. Row (342), parameter (number of labor personnel), unit (–), variable name (Nmsl)

$$\text{Formula} = \text{INT}\left(\frac{\left(\frac{264 \times \text{Wacs}}{6 \times 1000000}\right)^{0.4} \times 18}{1.4}\right)$$

$$= \text{INT}\left(\frac{\left(\frac{264 \times 120000}{6 \times 1000000}\right)^{0.4} \times 18}{1.4}\right) = 25$$

15. Row (343), parameter (O&M labor cost), unit (M$/a), variable name (Cmsl)

$$\text{Formula} = \frac{\text{Nmsl} \times \text{Sml}}{1{,}000{,}000} = \frac{25 \times 100{,}000}{1{,}000{,}000} = 2.5$$

16. Row (344), parameter (annual materials cost), unit (M$/a), variable name (Csmt)

$$\text{Formula} = \frac{(\text{cmm} \times \text{Fma} + \text{cmsp} \times \text{Fpp} + \text{cmcpr} + \text{cmcpo}) \times \text{Wpms}}{1{,}000{,}000}$$

$$= \frac{(0.06 \times 1.01 + 0.04 \times 1.012 + 0.03 + 0.03) \times 39{,}858{,}000}{1{,}000{,}000}$$

$$= 6.420$$

17. Row (345), parameter (annual insurance cost), unit (M$/a), variable name (Csins)

$$\text{Formula} = \left(\frac{\text{kmi}}{100}\right) \times \text{Cmscon} = \left(\frac{0.5}{100}\right) \times 121.45 = 0.607$$

18. Row (346), parameter (water plant O&M cost), unit (M$/a), variable name (Cmsom)

$$\text{Formula} = \text{Cmsm} + \text{Cmsl} + \text{Csmt} + \text{Csins}$$

$$= 0.6 + 2.5 + 6.42 + 0.61 = 10.13$$

12.1.9 Power plant

1. Row (394), parameter (levelized fixed charge rate), unit (%), variable name (lfc)

The annuity function in the interest rate and plant economic life as a parameter (resulting in the annual payment for an annuity per $ invested)

$$\text{Formula} = \text{PMT}\left(\frac{ir}{100}; \text{Lep}; -1\right) \times 100$$

$$\text{We use} = \frac{ir/100}{1-(1+ir/100)^{-n}} \times 100 = \frac{0.08}{1-(1+0.08)^{-10}} \times 100 = 14.9$$

12.1.10 Stand-alone reverse osmosis plant

1. Row (427), parameter (installed water plant production capacity), unit (m^3/d), variable name (Wacs1)

$$\text{Formula} = \text{Wacs} = 120{,}000$$

2. Row (428), parameter (annual average water production), unit (m^3/a), variable name (Wpms1)

$$\text{Formula} = \text{Wpms} = 39{,}858{,}000$$

3. Row (429), parameter (total construction cost), unit (M$), variable name (Cmscon1)

$$\text{Formula} = \text{Cmscon} = 121.45$$

4. Row (430), parameter (interest during construction), unit (M$), variable name (IDCs)

$$\text{Formula} = \text{Cmscon1} \times \left(\left(1+\frac{ir}{100}\right)^{\text{Lm}/24} - 1\right) = 121.45 \times \left(\left(1+\frac{8}{100}\right)^{24/24} - 1\right) = 9.716$$

5. Row (431), parameter (total investment), unit (M$), variable name (Cmsinv)

$$\text{Formula} = \text{Cmscon1} + \text{IDCs} = 121.45 + 9.72 = 131.17$$

6. Row (432), parameter (specific investment cost), unit [$/($m^3$/d)], variable name (csmsinv)

$$\text{Formula} = \frac{1{,}000{,}000 \times \text{Cmsinv}}{\text{Wacs1}} = \frac{1{,}000{,}000 \times 131.17}{120{,}000} = 1093$$

7. Row (433), parameter (annual water plant fixed charge), unit (M$/a), variable name (amsfc)

$$\text{Formula} = \frac{\text{Cmsinv} \times \text{lfc}}{100} = \frac{131.17 \times 14.9}{100} = 19.54$$

8. Row (434), parameter (annual water plant electric power cost), unit (M$/a), variable name (amsepc)

   The total stand-alone power use time times the combined power/water plant load factor times the number of hours per year times the levelized power cost. This term is ignored in the case of no power generation in the plant
9. Row (435), parameter (annual water plant purchased electric power cost), unit (M$/a), variable name (amsepu)

$$\text{Formula} = \frac{\text{Cpe} \times 1000 \times (\text{Apm} - \text{Acpm}) \times 8760 \times \text{Qms}}{1{,}000{,}000}$$
$$= \frac{0.06 \times 1000 \times (0.910 - 0) \times 8760 \times 26.77}{1{,}000{,}000} = 12.80$$

10. Row (436), parameter (annual water plant O&M cost), unit (M$/a), variable name (Cmsom1)

$$\text{Formula} = \text{Cmsom} = 10.13$$

11. Row (437), parameter (total annual required revenue), unit (M$/a), variable name (amsrev)

$$\text{Formula} = \text{amsfc} + \text{amsepc} + \text{amsepu} + \text{Cmsom1}$$
$$= 19.54 + 0 + 12.80 + 10.13 = 42.47$$

12. Row (438), parameter (total water cost), unit ($/m^3), variable name (Wmst)

$$\text{Formula} = \frac{1{,}000{,}000 \times \text{amsrev}}{\text{Wpms1}} = \frac{1{,}000{,}000 \times 42.47}{39{,}858{,}000} = 1.065$$

When the average management and labor salaries are changed to $100,000 and $50,000, respectively, the total water cost becomes 1.027 $/m^3. When the purchased electricity cost is changed to 0.05 $/kW(e)h, the total water cost becomes 1.012 $/m^3.

The entire procedure is summarized in Scheme 12.1

References

DEEP 5 User Manual, 2013. <https://www.iaea.org/sites/default/files/18/08/deep5-manual.pdf>.

Desalination Economic Evaluation Program (DEEP) User's Manual, September, 2000. Vienna: IAEA, IAEA-CMS-14, Printed by IAEA, Austria. <https://www-pub.iaea.org/MTCD/Publications/PDF/CMS-14.pdf>.
Desalination Economic Evaluation Program (DEEP) User's Manual (DEEP-3.0), April, 2006. Vienna: IAEA, IAEA-CMS-19, Printed by IAEA, Austria. <https://www-pub.iaea.org/MTCD/Publications/PDF/CMS-19_web.pdf>.

Appendix

Table A1 Some important parameters.

Avogadro number	6.023×10^{23} /mol
Boltzmann constant	1.381×10^{-23} J/K
Collision diameter of nitrogen	0.364 nm
Collision diameter of oxygen	0.346 nm
Collision diameter of water	0.264 nm
Dielectric constant of water	78.54
Diffusion constant of Na^+ in water	0.844×10^{-9} m^2/s
Diffusion constant of Cl^- in water	1.503×10^{-9} m^2/s
Diffusion constant of NaCl in water (0.05 mol/L, 18°C)	1.26×10^{-9} m^2/s
Electric charge	1.602×10^{-19} C
Radius of Na^+ ion	0.1840 nm
Radius of Cl^- ion	0.1207 nm
Gas constant	8.314 J/mol K

Table A2 Kinematic viscosity and density of water.

Temperature (°C)	Kinematic viscosity $\times 10^6$ (m^2/s)	Density (kg/m^3)
5	1.5182	1
10	1.3063	0.9997
15	1.1386	0.9991
20	1.0034	0.9982
25	0.8926	0.997
30	0.8007	0.9956
35	0.7234	0.994
40	0.6579	0.9922
45	0.6017	0.9902
50	0.5531	0.988
55	0.5109	0.9857
60	0.474	0.9832
65	0.4415	0.9806
70	0.4127	0.9778
75	0.3872	0.9748
80	0.3643	0.9718

Source: Anton Paar, https://wiki.anton-paar.com/ca-en/water/.

Table A3 Surface tension of water–air at different temperatures.

Temperature (°C)	Surface tension (N/m)
0	0.0757
10	0.0742
20	0.0728
30	0.0712
40	0.0696
50	0.0679
60	0.0662
70	0.0644
80	0.0626
90	0.0608
100	0.0588

Table A4 Data for the system NaCl-H_2O at 25°C.

Molality	Solute concentration (wt.%)	Osmotic pressure (bar)	Density of solution $\times 10^{-3}$ (kg/m^3)	Kinematic viscosity $\times 10^6$ (m^2/s)	Solute diffusivity (m^2/s $\times 10^9$, m^2/s)
0	0	0	0.9971	0.8963	1.610
0.1	0.5811	4.62	1.0011	0.9009	1.483
0.2	1.1555	9.17	1.0052	0.9054	1.475
0.3	1.7233	13.7	1.0091	0.9100	1.475
0.4	2.2846	19.5	1.0130	0.9147	1.475
0.5	2.8395	22.8	1.0169	0.9193	1.475
0.6	3.3882	27.4	1.0208	0.9242	1.475
0.7	3.9307	32.1	1.0248	0.9290	1.475
0.8	4.4671	36.8	1.0286	0.9338	1.477
0.9	4.9976	41.6	1.0322	0.9389	1.480
1.0	5.5222	46.4	1.0357	0.9440	1.483
1.2	6.5543	56.1	1.0427	0.9567	1.488
1.4	7.5640	66.1	1.0505	0.9685	1.492
1.6	8.5522	76.5	1.0581	0.9802	1.497
1.8	9.5194	87.0	1.0653	0.9923	1.505
2.0	10.4665	97.8	1.0722	1.0044	1.513

(*Continued*)

Molality	Solute concentration (wt.%)	Osmotic pressure (bar)	Density of solution × 10^{-3} (kg/m^3)	Kinematic viscosity × 10^6 (m^2/s)	Solute diffusivity ($m^2/s \times 10^9$, m^2/s)
2.2	11.3939	108.9	1.0790	1.0206	1.521
2.4	12.3022	120.3	1.0859	1.0365	1.530
2.6	13.1922	132.0	1.0927	1.0523	1.539
2.8	14.0642	144.0	1.0991	1.0683	1.548
3.0	14.9190	156.5	1.1056	1.0840	1.556
3.2	15.7568	169.1	1.1121	1.1047	1.565
3.4	16.5784	182.8	1.1185	1.1252	1.570
3.6	17.3840	195.4	1.1247	1.1457	1.575
3.8	18.1743	209.2	1.1309	1.1660	1.580
4.0	18.9496	223.3	1.1369	1.1862	1.585
4.2	19.7103	237.6	1.1429	1.2108	1.589
4.4	20.4569	252.4	1.1490	1.2350	1.594
4.6	21.1897	267.4	1.1550	1.2591	1.593
4.8	21.9092	283.0	1.1608	1.2832	1.593
5.0	22.6156	298.7	1.1666	1.3070	1.592
5.2	23.3093	315.0	1.1723		1.592
5.4	23.9908	331.4	1.1778		1.591
5.6	24.6602	348.5	1.1832		1.590
5.8	25.3179	365.7	1.1887		
6.0	25.9643	383.3	1.1941		

Table A5 Osmotic pressure of sea salt solutions at different temperatures (bar).

Temperature (°C)	Salts (wt.%)								
	1.0	2.0	3.45	5.0	7.5	10.0	15.0	20.0	25.0
25	7.2	14.5	25.4	38.0	60.1	85.1	147	233	355
40	7.5	15.1	26.6	39.8	63.2	90.2	155	243	365
60	7.6	15.9	28.0	42.0	66.8	95.2	164	253	385
80	8.2	16.6	29.3	43.9	69.7	99.2	170	263	395
100	8.5	17.1	30.3	45.3	72.0	102	175	274	405

Source: From Sourirajan, S., 1970. Reverse Osmosis. Academic Press, New York, NY.

Table A6 Osmotic pressure of seawater at various temperatures and concentrations of dissolved solids (bar).

Salt (wt.%)	Temperature (°C)										
	0	4.44	15.6	26.7	37.8	48.9	60	71.1	82.2	93.3	100
2	12.7	13.38	11.77	12.48	15.04	16.12	15.98	14.34	12.17	9.83	9.22
4	24.66	25.57	26.13	26.94	27.06	31.04	30.39	28.54	26.76	25.00	24.40
6	38.00	38.58	41.34	43.03	44.42	48.64	49.00	46.62	42.04	40.10	39.74
8	52.48	53.88	59.88	61.94	63.42	68.38	69.17	65.96	63.02	60.14	59.83
10	71.55	68.88	77.50	83.31	86.65	90.54	89.18	85.00	83.51	83.29	82.76
12	89.63	87.39	99.06	107.2	111.7	114.4	113.3	111.5	111.0	108.8	107.7
14	109.0	112.6	121.7	133.1	139.5	142.2	141.3	140.3	141.7	137.0	135.2
16	132.7	130.8	146.0	161.2	170.9	171.5	173.1	173.9	173.1	170.2	167.1
18	151.9	158.6	162.8	195.0	186.5	207.7	208.0	209.6	208.2	206.7	204.3
20	180.2	187.9	209.0	228.5	239.2	240.9	240.6	245.4	248.3	247.4	244.9
22	212.7	237.0	246.2	266.7	277.6	279.6	280.6	287.9	293.9	295.6	294.9
24	251.5	264.1	290.6	309.9	321.0	322.1	328.6	336.7	364.0	345.4	345.0
26	349.3	306.8	338.2	355.9	367.1	367.7	376.3	387.4	402.8	403.4	405.2

Source: From Sourirajan, S., 1970. Reverse Osmosis. Academic Press, New York, NY.

Table A7 Viscosity of gases.

Gas	Viscosity $\times 10^{-5}$ (Pa s) Temperature (°C)			
	0	20	50	100
Air	1.73	1.82	1.96	2.20
Ammonia	0.92	0.99	1.10	1.30
Argon	2.1	2.23	2.42	2.73
Benzene	0.7	0.75	0.81	0.94
Carbon dioxide	1.37	1.47	1.61	1.85
Carbon monoxide	1.66	1.74	1.88	2.10
Chlorine	1.23	1.32	1.45	1.69
Chloroform	0.94	1.01	1.11	1.28
Ethylene	0.97	1.03	1.12	1.28
Helium	1.87	1.96	2.10	2.32
Hydrogen	0.84	0.88	0.94	1.04
Methane	1.03	1.10	1.19	1.35
Neon	2.98	3.13	3.36	3.70
Nitrogen	1.66	1.76	1.89	2.12

(*Continued*)

Gas	Viscosity × 10^{-5} (Pa s) Temperature (°C)			
Nitrous oxide	1.37	1.47	1.61	1.84
Oxygen	1.95	2.04	2.18	2.44
Steam	0.92	0.97	1.06	1.24
Sulfur dioxide	1.16	1.26	1.40	1.64
Xenon	2.12	2.28	2.51	2.88

Source: The Engineering ToolBox, https://www.engineeringtoolbox.com/gases-absolute-dynamic-viscosity-d_1888.html.

Table A8 CO_2 permeability and CO_2/CH_4 permeability ratio for some polymers.

Polymer	CO_2 permeability (Barrer)	CO_2/CH_4 permeability ratio
PTMSP	33100	2.0
Silicone rubber	3200	3.4
Natural rubber	130	4.6
Polystyrene	11	8.5
Polyamide (Nylon 6)	0.16	11.2
Polyvinyl chloride	0.16	15.1
Polycarbonate (Lexan)	10.0	26.7
Polysulfone	4.4	30.0
Polyethylene terephthalate (Mylar)	0.14	31.6
Cellulose acetate	6.0	31.0
Polyetherimide (Ultem)	1.5	45.0
Polyethersulfone (Victrex)	3.4	50.0
Polyimide (Kapton)	0.2	64.0

Source: From Mulder, M., 1996. Basic Principles of Membrane Technology, second ed. Kluwer Academic Publisher B.V., The Netherlands.

Table A9 Gas permeability through polymer.

Polymer	Permeability (barrer)					
	CH_4	N_2	O_2	CO_2	H_2	He
Bisphenol-A polycarbonate	0.36	0.38	1.6	6.8	13.3	14
Bisphenol-A polysulfone	0.25	0.25	1.4	5.6	14	13
Tetramethyl bisphenol-A polycarbonate	0.89	1.1	5.6	18.6	–	46

(Continued)

Polymer	Permeability (barrer)					
	CH_4	N_2	O_2	CO_2	H_2	He
Tetramethyl bisphenol-A polysulfone	0.95	1.06	5.6	21	32	41
Tetrachloro bisphenol-A polycarbonate	0.22	0.36	2.3	6.7		27.4
Tetrabromo bisphenol-A polycarbonate	0.12	0.20	1.5	4.2		18
Hexafluoro bisphenol-A polycarbonate	1.05	1.7	6.9	24	43	60
Hexafluoro bisphenol-A polysulfone	0.55	0.67	3.4	12	27	33
Tetramethyl hexafluoro bisphenol-A polycarbonate	4.7	7.7	32	111		206
Tetramethyl hexafluoro bisphenol-A polysulfone	3.0	4.0	18	72	126	113
Tetrabromo hexafluoro bisphenol-A polycarbonate	0.89	1.8	9.7	32		100
Bisphenol-F polysulfone	0.19	0.20	1.1	4.5	10.6	10
Bisphenol-O polysulfone	0.18	0.196	1.1	4.3	10.4	10
Tetramethyl bisphenol-F polysulfone	0.58	0.61	3.3	5.5	15	29
Bisphenol polysulfone	0.25	0.24	1.3	5.6	14	12
Tetramethyl bisphenol polysulfone	1.27	1.21	5.8	31.8	28	36
1,5-dihydroxynaphthalene polysulfone	0.036	0.057	0.42	1.6	8.1	9.4
PMDA-ODA	0.059	0.10	0.61	2.7		8.0
PMDA-MDA	0.093	0.20	0.98	4.0		9.4
PMDA-IPDA	0.90	1.5	7.1	27		37.1
Bisphenol-P polysulfone	0.34	0.32	1.8	6.8	17	14
Tetramethyl bisphenol-P polysulfone	0.60	0.57	3.2	13.2	36	32
6FDA-ODA	0.38	0.83	4.34	23		51.5
6FDA-MDA	0.42	0.81	4.6	19		50
6FDA--IPDA	0.70	1.34	7.5	30		71.2
Dimethyl bisphenol-A polysulfone	0.07	0.091	0.64	2.1	11	12
Bisphenol of norbornane polycarbonate	0.48	0.47	2.4	9.1	22	19
Bisphenol-Z polycarbonate	0.092	0.105	0.60	2.2	9.1	10
Bisphenol of chloral polycarbonate	0.23	0.27	1.4	5.6	12.4	12
Hexamethyl bisphenol polysulfone	0.94	1,2	6.0	25.5	63	53
BTDA-6FpDA	0.155	0.31	1.9	7.3		
6FDA-6FpDA	1.6	3.4	16	64		137
Bisphenol-Z polysulfone	0.10	0.11	0.74	2.54	11.7	11.7
Dimethyl bisphenol-Z polysulfone	0.041	0.057	0.41	1.4	9.2	11
Polyester of bisphenol-A and terephthalic diacid chloride	0.68	0.625	2.5	11.6		16.8
Polyester of hexafluoro bisphenol-A and terephthalic diacid chloride	2.74	2.95	11.8	47.1		78.1
Polyester of 6,6′-dihydroxy-3,3,3′,3′-tetramethyl-1,1′-spirobiindane and terephthalic diacid chloride	4.2	2.98	13.4	55.7		68.0
Bisphenol-A polyetherimide	0.0356	0.0526	0.40	1.3		9.4
6,6′-Dihydroxy-3,3,3′,3′-tetramethyl-1,1′-spirobiindane polyetherimide	0.155	0.203	1.4	4.5		20.7
Polyester of bisphenol-A and 4,4′ biphenol diacid chloride	0.289	0.255	1.2	5.6		9.1
Polyester of bisphenol-A and 9H-fluorenone-2,7-dicarboxylic acid chloride	0.208	0.25	1.6	6.0		18.2
Polyester of hexafluoro bisphenol-A and 4,4′ biphenol diacid chloride	1.53	1.72	7.4	31.4		47.6
Polyester of hexafluoro bisphenol-A and 9H-fluorenone-2,7-dicarboxylic acid chloride	0.981	1.44	7.2	29.9		61.7

(Continued)

Polymer	Permeability (barrer)					
	CH_4	N_2	O_2	CO_2	H_2	He
12 hydrogen poly(ether ketone)	0.22	0.21	1.1	4.4	12.4	11.8
6 hydrogen 6 fluorine poly(ether ketone)	0.345	0.44	2.4	8.0	22.8	26.4
12 fluorine poly(ether ketone)	0.477	0.76	3.7	11.5	31.4	42.0
Poly(1,4-phenylene sulfide)	0.066	0.046	0.38	1.6		5.15
Poly(2,6-dimethyl-1,4-phenylene oxide)	4.1	3.5	14.6	65.5		82.3
Poly(2,6-diphenyl-1,4-phenylene oxide)	2.7	1.5	7.7	39.9		32.7
Bisphenol-A poly(bisketone)	0.14	0.13	0.75	3.3	8.9	8.9
Bisphenol-S poly(bisketone)	0.12	0.11	0.64	3.27	8.3	7.7
Bisphenol-A poly(bissulfone)	0.5	0.47	2.39	10.8	20.1	16.9
Polyarylate	0.40	0.33	1.7	7.5		14.0
Polyhydroxy ether	0.019	0.015	0.10	0.40		2.5
Poly(bisphenol-A tetrabutyl isophthalate)	1.43	1.17	5.95	24.2	41.9	34.3
PhTh/IA Poly(phenolphthalein isophthalate)	0.167	0.264	1.52	6.74	14.7	14.7
Poly(phenolphthalein tetrabutyl isophthalate)	1.16	1.06	5.6	23.8	41.3	35.6
Poly(fluorene bisphenol isophthalate)	0.616	0.556	3.03	10.3	26.1	22.3
Poly(fluorene bisphenol tetrabutyl isophthalate)	2.38	1.94	9.55	36.8	63.1	45.9
PhAnthr/IA Poly(phenolphthalein isophthalate)	0.334	0.342	2.05	9.0	19.7	18.0
Poly(tetrabromo bisphenol-A isophthalate)	0.142	0,178	1.29	4.93	17.8	16.3
Poly(tetrabromo bisphenol-A tetrabutyl isophthalate)	0.853	0.89	5.66	21.5	53.9	41.6
Poly(tetrabromo phenolphthalein isophthalate)	0.206	0.282	1.96	8.34	27.1	23.4
Poly(tetrabromo phenolphthalein tetrabutyl isophthalate)	1.09	1.19	7.41	30.6	67.4	51.3
Poly(tetrabromoflorene isophthalate)	0.567	0.704	4.83	20.4	62.2	38.5
Poly(tetrabromoflorene tetrabutyl isophthalate)	2.77	2.85	16.8	69.5	1.37	94.4
Poly(hexafluoro bisphenol-A tetrabutyl isophthalate)	3.47	3.63	15.7	56.9	86.7	91.1
Bisphenol acetophenone polycarbonate	0.419	0.361	1.84	9.48		13.9
Fluorene bisphenol polycarbonate	0.581	0.592	3.18	15.1		21.8
Bisphenol acetophenone polysulfone	0.32	0.278	1.56	8.12		13.3
Fluorene bisphenol polysulfone	0.531	0.484	2.76	13.8		21.3
6FDA-m-PDA	0.14	0.363	2.61	8.23	20.3	
6FDA-2,4-DATr	0.71	1.31	7.44	28.63	87.2	
6FDA-3,5-DBTF	0.45	1.17	6.43	21.64	58.6	
PMDA-4BDAF	0.36	0.66	2.9	11.8	24	
6FDA-4BDAF	0.51	0.98	5.5	19	46	
BPDA-6FpDA	0.761	1.21	6.65	27.4		
BTDA-6FmDA	0.014	0.047	0.39	1.05		
6FDA-6FmDA	0.08	0.261	1.8	5.1		
Bisphenol-M polysulfone	0.11	0.11	0.69	2.8	10.6	11.7
Tetramethyl bisphenol-M polysulfone	0.28	0.28	1.8	7.0	20	21
Polyester of bisphenol-A and isophthalic diacid chloride	0.27	0.28	1.5	5.6		15.1
Polyester of hexafluoro bisphenol-A and isophthalic diacid chloride	0.61	0.89	4.2	14.9		41.1
Polyester of 6,6′-dihydroxy-3,3,3′,3′-tetramethyl-1,1′-spirobiindane and isophthalic diacid chloride	1.14	1.0	5,3	18.9		39.1

(Continued)

Polymer	Permeability (barrer)					
	CH_4	N_2	O_2	CO_2	H_2	He
PMDA-3BDAf	0.17	0.29	1.4	6.12	19	
6FDA-3.3′-ODA	0.032	0.10	0.68	2.1	14	
6FDA-3BDAF	0.13	0.24	1.35	6.3	21	
6FDA-p-PDA	0.18	0.38	2.1	11.8	23	
PMDA-3,3′-ODA	0.0080	0.018	0.13	0.50	3.6	
Polyester of hexafluoro bisphenol-A isophthalate	0.799	1.11	5.23	19.1	37.3	47.8
3,4′-polysulfone	0.052	0.066	0.39	1.5	8.0	9.3
PMDA-mp′ODA	0.0258	0.0454	0.31	1.118	5.92	
PMDA-BATPHF	0.937	1.50	7.06	24.6	50.4	
BPDA-pp′ODA	0.0099			0.642	3.68	
BPDA-BAPHF	0.145	0.245	1.54	4.69	17.3	
BPDA-BATPHF	0.279	0.563	3.11	9.15	30.6	
BTDA-pp′ODA	0.0109	0.0236	0.191	0.625	4.79	
BTDA-BAPHF	0.105	0.195	1.14	4.37	16.1	
BTDA-BATPHF	0.189	0.370	2.17	6.94	24.6	
6FDA-mp′DA	0.125	0.259	1.57	6.11	23.7	
6FDA-APAP	0.217	0.473	2.89	10.7	38.2	
6FDA-BATPHF	0.703	1.30	6.50	22.8	55.4	
6FDA-DAF	0.63	1.27	7.85	32.2		98.5
Poly(amide amino acid) of 2,2- and bis(3,4-dicarboxyphenyl) hexafluoropropane dianhydride tetraaminodiphenyl ether	0.087	0.17	0.97	3.69		18.9
Polypyrrolone of 2,2- bis(3,4-dicarboxyphenyl) hexafluoropropane dianhydride tetraaminodiphenyl ether	0.54	1.2	7.9	27.6		89.0

Source: Park, J.Y., Paul, D.R., 1997. Correlation and prediction of gas permeability in glassy polymer membrane materials via a modified free volume based group contribution method. J. Membr. Sci. 125, 23–39.

Table A10 Contact angle of water for different polymers.

Polymer	Contact angle (in degrees)
Polyvinyl alcohol[a]	51
Polyvinyl acetate[a]	60.6
Nylon 6[a]	62.6
Polyethylene oxide[a]	63
Nylon 6,6[a]	68.3
Nylon 7,7[a]	70
Polysulfone[a]	70.5
Polymethyl methacrylate[a]	70.9
Nylon 12[a]	72.4

(Continued)

Polymer	Contact angle (in degrees)
Polyethylene terephthalate[a]	72.5
Epoxides[a]	76.3
Polyoximethylene[a]	76.8
Polyvinylidene chloride[a]	80
Polyphenylene sulfide[a]	80.3
Acrylonitrile butadiene stylene[a]	80.9
Nylon 11[a]	82
Polycarbonate[a]	82
Polyvinyl fluoride[a]	84.5
Polyvinyl chloride[a]	85.6
Nylon 8,8[a]	86
Nylon 9,9[a]	86
Polystyrene[a]	87.4
Polyvinylidene fluoride[a]	89
Poly n-butyl methacrylate[a]	91
Polytrifluoroethylene[a]	92
Nylon 10,10[a]	94
Polybutadiene[a]	96
Polyethylene[a]	96
Polychlorotrifluoroethylene[a]	99.3
Polypropylene[a]	102.1
Polydimethylsiloxane[a]	107.2
Poly t-butyl mathacrylate[a]	108.1
Fluorinated ethylene propylene[a]	108.5
Hexatriacontane[a]	108.5
Paraffin[a]	108.9
Polytetrafluoroethylene[a]	109.2
Poly(hexafluoropropylene)[a]	112
Polyisobutylene[a]	112.1

[a]Diversified Enterprises, https://www.accudynetest.com/polytable_03.html?sortby = contact_angle.

Table A11 Saturation vapor pressure and heat of vaporization of water.

Temperature (°C)	Saturation vapor pressure (kPa)	Heat of vaporization (kJ/kg)
0	0.611	2501.7
5	0.872	2489.7
10	1.227	2477.9
15	1.704	2466.1
20	2.337	2454.3

(Continued)

Temperature (°C)	Saturation vapor pressure (kPa)	Heat of vaporization (kJ/kg)
25	3.166	2442.5
30	4.241	2430.7
35	5.622	2418.8
40	7.375	2406.9
45	9.582	2394.9
50	12.34	2382.9
55	15.74	2370.8
60	19.92	2358.6
65	25.01	2346.3
70	31.16	2334.0
75	38.55	2321.5
80	47.36	2308.8
85	57.80	2296.1
90	70.11	2283.2
95	84.53	2270.2
100	101.33	2256.9

Source: Smith, J.M., Van Ness, H.C., Abbott, M.M., 1996. Introduction to Chemical Engineering Thermodynamics, 5th ed. McGraw-Hill.

Table A12 Thermal conductivity of gases (at 100 kPa or at saturation vapor pressure when it is below 100 kPa).

Gas	Thermal conductivity × 10^3 (W/m K) Temperature (K)	
	300	400
Air	26.2	33.3
Argon	17.9	22.6
Boron trifluoride	19.0	24.6
Hydrogen (low pressure)	186.9	230.4
Sulfur hexafluoride (low pressure)	13.0	20.6
Water	18.7	27.1
Hydrogen sulfide	14.6	20.5
Ammonia	24.4	37.4
Helium (low pressure)	156.7	190.6
Krypton (low pressure)	9.5	12.3
Nitric oxide	25.9	33.1
Nitrogen	26.0	32.3
Nitrous oxide	17.4	26.0
Neon (low pressure)	49.8	60.3

(Continued)

Gas	Thermal conductivity × 10^3 (W/m K) Temperature (K)	
Oxygen	26.3	33.7
Sulfur dioxide	9.6	14.3
Xenon (low pressure)	5.5	7.3
Dichlorodifluoromethane	9.9	15.0
Tetrafluoromethane (low pressure)	16.0	24.1
Carbon monoxide (low pressure)	25.0	32.3
Carbon dioxide	16.8	25.1
Trichloromethane	7.5	11.1
Methane	34.1	49.1
Methanol	—	26.2
Acetylene	21.4	33.3
Ethylene	20.5	34.6
Ethane	21.3	35.4
Ethanol	14.4	25.8
Acetone	11.5	20.2
Propane	18.0	30.6
Butane	16.4	28.4
Pentane	14.4	24.9
Hexane	—	23.4

Source: Engineer's Edge, https://www.engineersedge.com/heat_transfer/thermal-conductivity-gases.htm.

Table A13 Thermal conductivity of polymers.

Polymer	Thermal conductivity (W/m K)
Acrylonitrile butadiene styrene	0.130–0.190
Acrylonitrile styrene acrylate	0.170
Acrylonitrile styrene acrylate/polycarbonate blend	0.170
Cellulose acetate	0.250
Cellulose acetate butyrate	0.250
Cellulose propionate	0.190
Chlorinated polyvinyl chloride	0.160
ECTFE	0.150
Ethylene vinyl alcohol	0.340–0.360
Fluorinated ethylene propylene	0.250
High-density polyethylene	0.45–0.50
High-impact polystyrene	0.110–0.0.140

(*Continued*)

Polymer	Thermal conductivity (W/m K)
Ethylene methyl acrylate copolymer	0.230–0.250
Low-density polyethylene	0.320–0.350
Linear low-density polyethylene	0.350–0.450
Polyamide (PA) 46	0.300
PA 6	0.240
PA 6-10	0.210
PA 66	0.250
PA 66 impact modified	0.240–0.450
Polyamideimide	0.240–0.540
Polyarylate	0.180–0.210
Polybutylene terephthalate	0.210
Polycarbonate, high heat	0.210
Polyetheretherkotone	0.250
Polyetherimide	0.220–0.250
Polyetherketoneketone	1.750
Polyethersulfone	0.170–0.190
Polyethylene terephthalate	0.290
Polyethylene terephthalate glycol	0.190
Perfluoroalkoxy	0.190–0.260
Polyimide	0.100–0.350
Polylactide	0.100–0.195
Polymethylmethacrylate/acrylic (PMMA/acrylic)	0.150–0.250
PMMA/acrylic impact modified	0.200–0.220
Polyoxymethylene (acetal)	0.310–0.370
Polypropylene copolymer	0.150–0.210
Polypropylene homopolymer	0.150–0.210
Polyphenylene ether	0.160–0.220
Polyphenylene sulfide	0.2900.320
Polysulfone	0.120–0.260
Polytetrafluoroethylene	0.240
Polyvinyl chloride (PVC), plasticized	0.160
PVC rigid	0.160
Polyvinylidene chloride	0.160–0.200
Polyvinylidene fluoride	0.180
Styrene acrylonitrile	0.150
Stylene maleic anhydride	0.170

Source: OMNEXUS, https://omnexus.specialchem.com/polymer-properties/properties/thermal-insulation https://omnexus.specialchem.com/polymer-properties/properties/thermal-insulation https://omnexus.specialchem.com/polymer-properties/properties/thermal-insulation.

Index

Note: Page numbers followed by "*f*" and "*t*" refer to figures and tables, respectively.

A

Active layer (AL), 51
  AL-facing-DS, 51
  on top of porous sublayer, 87*f*
Active layer–facing-feed water (AL-facing-FW), 51
Activity coefficient, 115, 119
  of ethanol/water mixture, 119
  in membrane, 116
Adsorption, 155, 158–159
Air gap membrane MD (AGMD), 133–134
Arrhenius relationship, 78

B

Binary gas mixture separation, 84–86
Binodial line, 22–23
Bioreactor modeling, 218–227
Boundary concentration, 36–37
Boundary conditions (BC), 80–81, 169–170, 221
Breakthrough curve, 157
Brownian diffusion, 71–73
  coefficient, 73
  flux obtained by equations for, 74*t*

C

Carman–Kozeny equation (C-K equation), 157
  for flux calculation, 157
Cascade and recycle, 193–195
CO_2 permeability and CO_2/CH_4 permeability ratio for polymers, 253*t*
Cohesive energy, 1–2
  density, 1–2
Concentration polarization, 35–37
  in forward osmosis, 51–52
  prediction of RO performance, 37–42
Contact angle of water for different polymers, 256*t*
Continuous-type bioreactor system, 216
Cost of water, 233–246
Cross-flow types, 193

D

Darcy's law, 157
Desalination, 51, 168
Desalination Economic Evaluation Program (DEEP), 233–246
Diffusivity (*D*), 83, 121
  of NaCl, 67
  selectivity, 84
Direct contact MD (DCMD), 133–134
  transport in, 135–144
    heat transfer, 135–138
    mass transfer, 138–144
Draw solution (DS), 49–50, 210–211
Dry–wet phase inversion process, 17

E

Economic parameters input data, 236–237
Emergency bag, 50
Evaporative cooling water make-up, 51
External concentration polarization, 51

F

Fick's first law, 77, 81
Fick's second law, 79–80
Flat-sheet membrane for RO, 176–183
Flory–Higgins equation, 18
  solutions for ternary system using, 23–29
Flux paradox for colloidal suspension, 73
Forward osmosis (FO), 32*f*, 49–58. *See also* Reverse osmosis (RO)
  applications of, 50–51
  concentration polarization in, 51–52
  principles of, 49–50
  transport, 52–58
Forward osmosis–reverse osmosis hybrid system (FO–RO hybrid system), 51, 210–215

G

Gas mixture separation, 98–102
Gas permeability through polymer, 253*t*
Gas separation, 183–189
  membranes, 148
Gas separator
  cascade, 203–206
  recycle, 206–210
Gas transport in porous membrane, 96–104
  measurement of pore size, 102–104
  pore size distribution, 102–104

Gas transport in porous membrane (*Continued*)
  separation of gas mixture, 98–102
  transport mechanism, 96–98
Gel concentration, 69
Gel model, 69–70
Gibbs adsorption isotherm, 42
Gibbs–Duhem relation, 20
Graetz number (*Gr*), 152

H

Heat transfer, 135–138
Henry's law, 77
Hildebrand's equation, 1–2
Hollow fiber
  for gas separation, 183, 184*f*
  modules for reverse osmosis, 165–175
Hybrid systems, 195
Hydrogen bonding force component, 2–3
Hyperfiltration, 61

I

Inertia lift, 73–74
Initial conditions (IC), 80
Internal concentration polarization, 51
Internal pressure, 1–2
International Atomic Energy Agency (IAEA), 233

K

Kimura–Sourirajna's equations, 37
Kinematic viscosity and density of water, 249*t*
Knudsen diffusion, 150. *See also* Brownian diffusion
Knudsen flow, 97–98
Knudsen model, 138
Knudsen number (K_n), 138

L

L'Hopital's rule, 169–170
Liquid entry pressure of water, 148–149

M

Mass balance in membrane adsorption, 157–161
Mass transfer, 138–144
  for nonwetted gas–liquid membrane contactor, 149*f*
  resistances, 134
Membrane absorption. *See* Membrane contactor
Membrane adsorption
  C-K equation, 157
  equipment for, 156*f*
  mass balance in, 157–161
  principle of, 156*f*
  process outline, 155–161
Membrane bioreactor, 216–217, 217*f*
Membrane contactor, 147–154
  device, 151*f*
  transport in, 149–152
  Wilson plot, 152–154
Membrane distillation (MD), 133
  configurations of, 134–135, 135*f*
  process principles, 133
  transport in DCMD, 135–144
Membrane gas separation
  gas transport in porous membrane, 96–104
  mixed matrix membrane, 105–109
  solution-diffusion model, 77–96
Membrane module
  gas separation, 183–189
  reverse osmosis, 165–183
Membrane reactor, 215–227
  bioreactor modeling, 218–227
  membrane bioreactor, 216–217, 217*f*
Membrane system
  flow types, 193–195, 194*f*
    cascade and recycle, 193–195
    cross-flow types, 193
    hybrid systems, 195
  FO–RO hybrid system, 210–215
  gas separator systems, 203–210
  membrane reactor, 215–227
  RO systems, 196–203
Membrane water performance data, 235–236
Metastable region, 19
Microfiltration (MF), 71–74, 133
Mixed matrix membrane (MMM), 105–109
Molecular diffusion, 97
Monovalent electrolytes, 63

N

Nanofiltration (NF), 61, 133, 196
  monovalent electrolytes, 63
  pore model for, 62*f*
  solution in general, 61–63
  solution method, 63–67
Nernst–Planck equation, 61

O

Ohm's law, 86–87
Ordinary-diffusion model, 138
Osmosis, 31
Osmotic pressure, 34
  of sea salt solutions at different temperatures, 251*t*
  of seawater at temperatures and concentrations of dissolved solids, 252*t*
Overall mass transfer coefficient, 150–151

P

Permeability (*P*), 77–78, 83, 85, 93
Pervaporation
  transport, 113–121
    model by Greenlaw and coworkers, 121–125
    new model for, 125–130
Phase-inversion technique, 17

Plait point, 20
Polymer solution thermodynamics, 17–23
Pore models, 42–49
  Glückauf model, 45–49
  for nanofiltration, 62*f*
  physicochemical data pertinent, 44*t*
  preferential sorption-capillary flow model, 42–44
Pore size
  distribution, 102–104
  measurement of, 102–104
Porous membrane
  gas transport in, 96–104
  measurement of pore size, 102–104
  pore size distribution, 102–104
  separation of gas mixture, 98–102
  transport mechanism, 96–98
Power plant, 245
Pressure-retarded osmosis (PRO), 32*f*, 49–58
  principles of, 49–50

R

Recycle
  cascade and, 193–195
  gas separator, 206–210
  reverse osmosis, 200–203
Resistance model, 86–96
Reverse osmosis (RO), 31–34, 32*f*, 61, 133, 165, 195–203, 233. *See also* Forward osmosis (FO)
  considering concentration polarization, 37–42
  flat-sheet membrane, 176–183
  hollow fiber modules, 165–175
  nanofiltration cascade, 196–200
  osmotic pressure data, 39*t*
  plant cost input data, 237–238
  principles of, 49–50
  recycle, 200–203
Reverse osmosis–nanofiltration cascade (RO–NF cascade), 196–200
Reynolds number (*Re*), 151–152

S

Saturation vapor pressure and heat of vaporization of water, 257*t*
Schmidt number (*Sc*), 151–152
Selectivity of gas, 84
Separation
  of binary gas mixture, 84–86
  factor, 85, 117–118
  of gas mixture, 98–102
Shear-induced diffusion, 73
Sherwood number (*Sh*), 151–152
Simulation algorithm, 172
Single-purpose plant performance, 238
Sodium chloride (NaCl), 250*t*
  concentration, 43–44
  data for NaCl-H_2O system, 250*t*
Solubility
  parameter, 1
    examples of, 3–14
    group contribution to, 7*t*
    of polymers, 14*t*
    of solvent, 13*t*
    theory, 1–3
  selectivity, 84
Solubility coefficient (*S*), 78, 83
Solute gas, 149–150
Solution method, 63–67
Solution-diffusion model (S-D model), 31, 77–96
  resistance model, 86–96
  separation of binary gas mixture, 84–86
  steady-state transport, 77–79
  unsteady-state evaluation of *S* and *D*, 79–83
Solvent/non-solvent exchange, 29
Spinodial line, 22–23
Spiral-wound module, 176–177
Stand-alone reverse osmosis water plant, 245–246
  costs, 242–244
  performance, 238–242
Steady-state transport, 77–79
Sulfonated polyphenylene oxide (SPPO), 79
Support layer (SL), 51
Surface tension of water–air at different temperatures, 250*t*
Sweep gas membrane distillation (SGMD), 135

T

Technical parameter input data, 235
Ternary system, solutions for, 23–29
Thermal conductivity
  of gases, 258*t*
  of polymers, 259*t*
Tie line, 22–23
Time-lag method, 79–83
Triangular phase diagram
  solutions for ternary system, 23–29
  thermodynamics of polymer solution, 17–23

U

Ultrafiltration (UF), 69–70, 133, 215–216
University of California Los Angeles (UCLA), 42
Unstable region, 19
Unsteady-state evaluation of *S* and *D*, 79–83

V

Vacuum membrane distillation (VMD), 134–135
Viscosity of gases, 252*t*
Viscous flow, 99
Viscous model, 138

W

Water permeation coefficient, 31

Water production cost
calculation, 233–246
case identification, 235
economic parameters input data, 236–237
membrane water performance data, 235–236
power plant, 245
recovery ratio of spiral wound module, 236*t*
reverse osmosis plant cost input data, 237–238
single-purpose plant performance, 238
site characteristics box, 235
stand-alone reverse osmosis water plant
costs, 242–244
performance, 238–242
stand-alone reverse osmosis plant, 245–246
technical parameter input data, 235
Wheatstone bridge model, 90
Wilson plot, 152–154

Printed in the United States
by Baker & Taylor Publisher Services